TELESCOPES FOR SKYGAZING

By Henry E. Paul, Ph.D.

THIRD EDITION

AMPHOTO

American Photographic Book Publishing Co., Inc.
Garden City, N.Y. 11530

Published in Garden City, New York, by American Photographic Book Publishing Co., Inc.

Library of Congress Catalog Card Number: 65-26425

ISBN: 0-8174-2408-3

Manufactured in the United States of America

PREFACE

A curious boy fingered through the scientific books on the shelf of the Carnegie Library in Guthrie, Oklahoma, not really knowing what he was searching for. Pulling down a cracker-thin red book unpretentiously marked *Amateur Telescope Making,* he was soon captured by the uniquely clear and informative pencil sketches by Russell Porter, and the encouraging style of the editor, A. G. Ingalls. These enthusiastic and persuasive individuals pioneered in showing how *anyone* who persevered could make the all-important light collector of an astronomical reflecting telescope, its mirror, from a chunk of glass, grinding grit, polishing material and a home silvering method; and that such a mirror in a simple wooden tube could rival expensive lens telescopes so elegantly housed by the professional astronomers.

Soon thereafter your author, as an amateur astronomer, was proudly displaying a crude telescope and excitedly pointing out lunar mountains to his companions and hearing their startled exclamations at first observing Jupiter's satellites and Saturn's rings. Like most of you, I've had many hobbies, but none have become as firmly entrenched or survived as long as that of being an amateur observer and telescope builder.

Now, if you have never explored through a telescope the many astronomical wonders shining in the night sky and pointed these out to your friends, you have a thrilling experience waiting. This book really came about because of a sincere wish to encourage all scientifically inclined individuals to obtain a simple astronomical telescope and join the rapidly growing throng of space-minded amateur astronomers and their clubs. Everyone can enjoy amateur skygazing, and the telescope can be as modest as you wish. Build your own! I've always felt my first small investment gave an unmatched return in the pleasures of freely exploring the wonderful new world of outer space.

When you own a telescope, the night sky offers you, your family, and friends the means for a leisurely armchair tour to the moon, the planets, or the challenging, dim, and mysterious island universes right from your own back yard. By no means overlook the educational possibilities for youngsters or the new friends you can make through such an instrument.

At this time I'd like to express my appreciation to my pen friends, to the enthusiastic observers, builders and photographers I know at the Rochester Academy of Science, to the Stellafane group, and most especially to personal friends, such as Ben Cleveland, Ralph Dakin, Horace Dall, Paul Davis, Jim Gagan, Alan Gee, Stan Gibson, George Keene, Alan McClure, Walt Semerau, Jack Smith, P. Van Newland, George Hageman and Dr. A. Neill, who have made this hobby even more pleasurable.

Fig. 2

ACKNOWLEDGMENT

The author is most appreciative for the assistance of many of the outstanding professional builders of optical instruments for the amateur in providing illustrative material and information. I am particularly indebted to Mrs. L. Braymer (Questar), Charles Brisley (Star-Liner), Dave Bushnell, Tom Cave, Norman Edmund, Lawrence Fine (Unitron), Tom Johnson (Celestron-Pacific), John Krewalk (Criterion), and to the observatories, publishers and outstanding photographers whose photos are reproduced here.

The fine small instruments illustrated above (Bushnell), the excellent 4-inch refracting telescope (Unitron) at lower left and the compact 8-inch deluxe reflector (Cave) at lower right are three representative examples of the excellent equipment and size ranges now available to the amateur.

I sincerely appreciate the availability of outstanding amateur work that provided most excellent photo material, and particularly wish to thank Ralph Dakin, Paul W. Davis, Charles Spoelhof and my editors for helpful advice, and M.L.O. for manuscript preparation.

Fig. 3

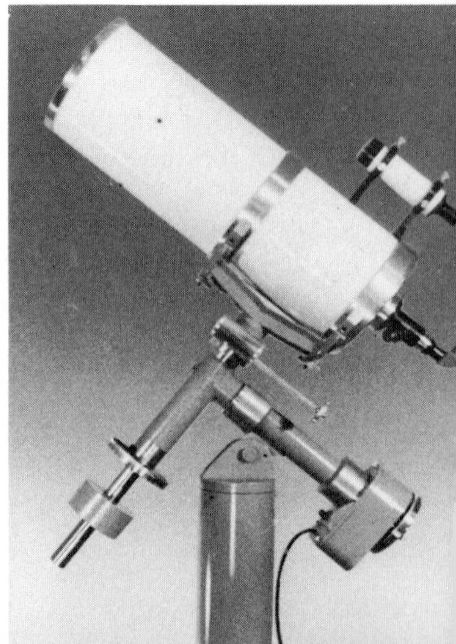

TABLE OF CONTENTS

Fig. 1 — **In this space-exploring age more and more telescopes are being trained on the many objects that gleam in the night sky. A good astronomical telescope lets you survey our moon, study the sun and neighboring planets, or search for a new comet. It even permits looking far out into space at mysterious glowing nebulae and spiraling galaxies of stars that populate the dim outer regions far beyond our own Milky Way system.**

1

EXPLORING OUTER SPACE

Few thrills exceed that of exploring the endless expanse of outer space through the magnifying eye of a powerful telescope optically spanning the great distances between us and our interesting neighbors, the sun, moon and planets. Astronomy has made its greatest progress with the aid of the telescope. The universal interest in astronomy over the years is strikingly illustrated by the astronomical stamps issued by many nations that honor famous astronomers and depict outstanding astronomical subjects. Accordingly, a select few of these are used to introduce some of the striking sky objects we are soon to see.

Just imagine Galileo's excitement when he first observed a strange procession of tiny moons swinging about their giant banded planet, Jupiter,

1.1 – First closer views of the mysterious objects of outer space have always thrilled man. (left) **Galileo excitedly points out the wonders of his new telescope on this gray-green 1942 Italian commemorative stamp.** (right) **Our first earth orbiting flight by Lt. Col. John H. Glenn, Jr., in 1962 is honored by this blue and gold U. S. stamp.**

and when he pointed out the many wonders of his telescope to the nobility, as depicted in Fig. 1.1. Our astronauts too must have marveled at their first views of jet-black, star-studded daytime skies, the earth's colorful sharp edge and our sun as a blazing, nearby star.

During many observing parties, at which I have served to adjust telescopes, guide eyes to eyepieces in the dark, and keep children glued to stepladders, I am constantly impressed by the startled exclamations of delight from new observers at their first sight of the gleaming planets and satellites that thrilled Galileo so long ago.

With the precision telescopes you can buy, assemble or build, the myriad of fascinating objects that glow, gleam and sparkle high in our night sky are now accessible from your back yard. Even the smallest of modern quality telescopes increases your seeing power fifty to a hundred times — enough for you to see far more than did the famous early astronomers. However, before presenting the many instruments currently available, I'd like first to take you on a brief tour through the night skies by means of a telescope's exploring lenses. Many of the astrophotos displayed were made by amateurs, and are used both to illustrate the fine work amateurs can accomplish and to present views that more nearly match what can actually be seen with amateurs' telescopes.

THE SUN

The nearest star, our sun, has awed man since time immemorial, and the ancients believed a chariot transported it across the sky. We now know it to be our prime source of light and energy. Periodically the sun puts on a spectacular show when our moon comes between it and the earth, a striking event at any date. Stamps depict these events. Even a partial eclipse presents an interesting sequence to the camera, as shown at the top of Fig. 1.2. During rarer annular, or total, eclipses we may see some of the phenomena shown next, a gleaming rim of light, the spectacular "diamond ring" effect and the glowing coronal halo at totality streaming out for millions of miles into the darkened sky. At this time prominences like weird tongues of red flame shoot out from the sun's edge, sometimes settling back to form loops as illustrated.

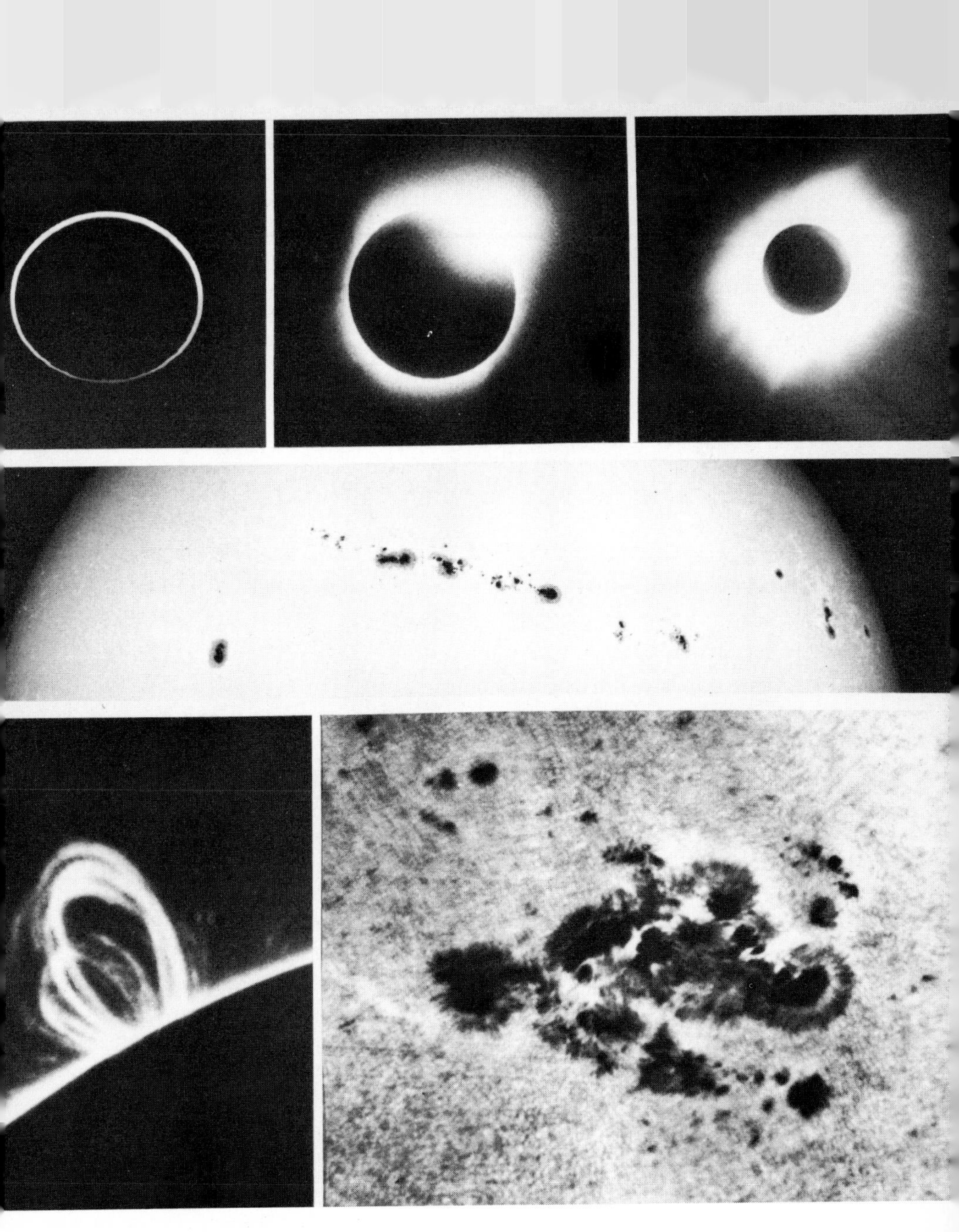

1.2 — **The sun is always a majestic object. It stages spectacular shows during total eclipses and periods of high sunspot activity. Partial eclipse by author; annular by R. Dakin; diamond ring by P. W. Davis; totality by F. Laudenklos; sunspot chain by Hans Arber; prominence by J. Klepesta; giant sunspot by R. and D. Davis.**

Even on ordinary days the sun can present an interesting array of dark sunspots or freckling, except for a few "quiet" years of low solar activity. In the figure we see a long string of immense spots as photographed on the day of the highest recorded sunspot activity. Below this is a detailed closeup of the sun's "rice grain" surface structure around a giant spot that could easily swallow our earth.

As fascinating as our sun may be, and as familiar as it is to us all, it strangely presents one of the few dangers to the unwary amateur observer. One should *never* look at the sun directly or through any astronomical or optical instrument without simple but proper filtering means readily available to reduce its intense light and heat, which could otherwise irreparably damage the retina of the eye.

OUR MOON

We can't all go to the moon, but we can easily get a good close look at it, and telescopes and rockets are busily doing so. In Fig. 1.3 we see exactly how the moon appears at fifty times magnification, a power easily reached by the smallest instrument worthy of being called an astronomical telescope. Hold the book 15 inches away for proper perspective. Now note in the insert (lower right corner) how small the moon appears to the unaided eye — actually a pencil eraser held at arm's length easily covers the moon as viewed in the sky. Most astronomical telescopes reverse the image as shown.

Note how only 50-times magnification clearly shows the massive jagged mountains, as high as any on earth, and the long ranges, bright rays and shadowy flat plains. Yet the amateur who owns a medium-sized instrument, as the widely popular 6-in. reflector telescope, quite often uses powers in the range of 300 times magnification. Let's now turn such a high-power telescope to the Mare Imbrium area, as marked near the bottom in Fig. 1.3, for the 250-power view shown in Fig. 1.4. The added detail in peaks, valleys and plains is striking indeed. Many areas of the moon appear even more

1.3 – The gibbous moon at last quarter, exactly as seen through a 50-power astronomical telescope (which inverts the image). Naked eye view depicted lower right. Arrows mark areas shown in closeups of the next figure and Fig. 10.7b. Photo by T. Osypowski.

1.4 – **The 83-mile-long great lunar Alpine Valley at lower left, Plato's smooth gray floor at the right, and the 7000 ft. isolated spire of Piton above pointing its fingerlike shadow at the Cassini crater to its left are clearly shown on this excellent 250-power closeup by T. Osypowski.**

1.5 – **Two informative views show direct angular size comparisons of our earth-lit moon and Saturn at lower magnification** (left) **and Mars near our moon's edge** (right) **at higher power, emphasizing the great distance of these major planets. Saturn conjunction photo by Yerkes Observatory and Mars pre-occultation photo by Lowell Observatory.**

rugged and challenging. One may now observe strange, spidery, trenchlike clefts, sometimes interrupted with craterlets and peculiar walls, ridges, cliffs, and rills. Occasionally an isolated mountain peak along the terminator rears up majestically to cast its long, fingerlike black shadow, as does the peak Piton in Mare Imbrium. See Fig. 10.7b for the Clavius area marked near the top of Fig. 1.3 and for two other closeup views taken during first quarter. Occasionally the moon occults a planet, presenting a thrilling view, and showing in a striking way how distance changes the apparent size relationship between the giant planet Saturn, distant Mars and our relatively tiny but nearby moon (Fig. 1.5).

Our earth's atmosphere plus distance severely limits a telescope's ability to see refined detail on the moon. Space platforms eliminate some of these problems, as illustrated by photos from Rangers VII and VIII presented in Fig. 1.6. Recent space photos provide more detail than all the observatories in the world.

NEIGHBORING PLANETS

Nine planets, including the Earth, swing steadily around our sun in ceaseless orbit. Copernicus fixed their relationship to the sun, and Kepler formulated the laws governing their motion in the sky. Now great observatories devote extensive programs to planetary study, as Pic du Midi high in the French Pyrenees (above) and our own Lowell Observatory.

Mercury is nearest the blazing sun, then comes Venus, followed by the Earth. Quite a journey separates us from Mars, the neighboring planet beyond us that is most likely to have some form of life as we know it. We then move on out to the giant planets, impressive clouded-banded Jupiter, and Saturn with its spectacular ring system. Each giant is surrounded by a covey of whirling moons. Finally Uranus, Neptune and Pluto swing slowly about the sun in their distant cold and lonely orbits.

Four of our neighboring planets, Venus, Mars, Jupiter and Saturn, are visible at regular intervals in the night sky to all amateurs' telescopes and present ever-changing views. Mercury and Venus go through relatively rapid phase changes like miniature moons, as shown in Fig. 1.7. Venus is a bright

1.6 – **A view of the moon by Jet Propulsion Lab's Ranger VIII camera shows the southwest corner of the Sea of Tranquillity from 151 miles. Insert shows strange rocklike mass as photographed from only about three miles by the previous Ranger before striking in the Sea of Clouds. (See color section for "space" photos.) Photos courtesy National Aeronautics and Space Administration.**

and easy object for lower-power telescopes, and in its crescent phase always pleases first viewers. Mercury, being near the sun, is only glimpsed occasionally before sunrise or after sunset, and at rare intervals transits as a black dot across the face of the sun.

Outside our earth's orbit swings Mars, the "red planet," the God of War to many ancients. It's a fickle neighbor, sometimes approaching close for favorable views, as shown at the left in upper Fig. 1.8, where we may see its snow-white polar cap and strong reddish and delicate green surface colorings, which change with the planet's seasons; and then receding rapidly to the other side of the sun, growing so small as to be a disappointing subject for even the largest instruments. When in favorable position Mars is a delicate object of beauty in precision-made medium-sized amateur's telescopes.

The giant of all planets, Jupiter, has a volume over a thousand times that of our earth. The cloud-like bands covering its whirling surface change rapidly in color and structure, and a great red spot, as shown in lower Fig. 1.8, develops as an elliptical patch to sweep slowly across one hemisphere, dramatically indicating planetary surface rotation. Four of Jupiter's satellites or moons are easily seen shining like stars — even with a pair of binoculars. At the higher magnifications of most amateurs' scopes this train of satellites presents an impressive, ever-changing sight as it swings about the massive mother planet, passing behind it to be both eclipsed and occulted, and then swinging in front of it to cast shadows as sharp jet-black dots on the bright cloud surface.

The great ringed planet Saturn always elicits gasps of astonishment and admiration when first observed by the amateur. It has been justly selected

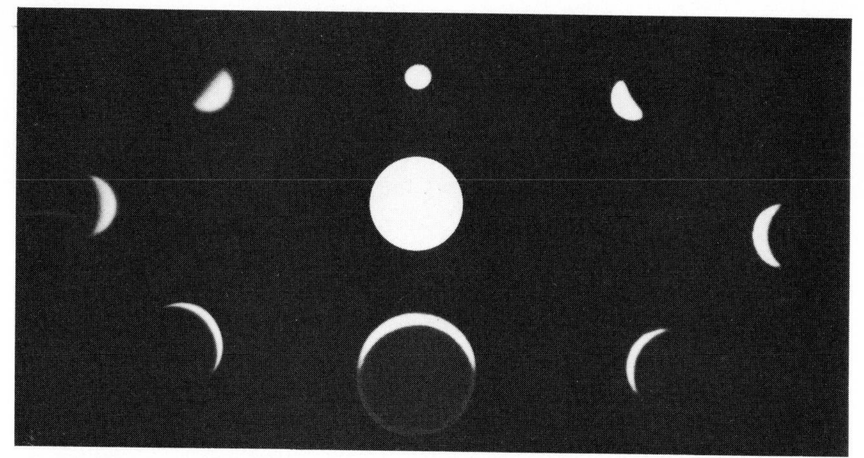

1.7 – **Venus on the far side of the sun** (top) **grows markedly in size to become a crescent as it swings around toward us to approach its nearest point** (below), **reversing to a morning crescent to rapidly recede in its solar circuit. Left mid-phases by Hans Pfleumer, right mid-phases by G. A. T. Heillegger, two extreme phases courtesy Lowell Observatory – pictorially arranged.**

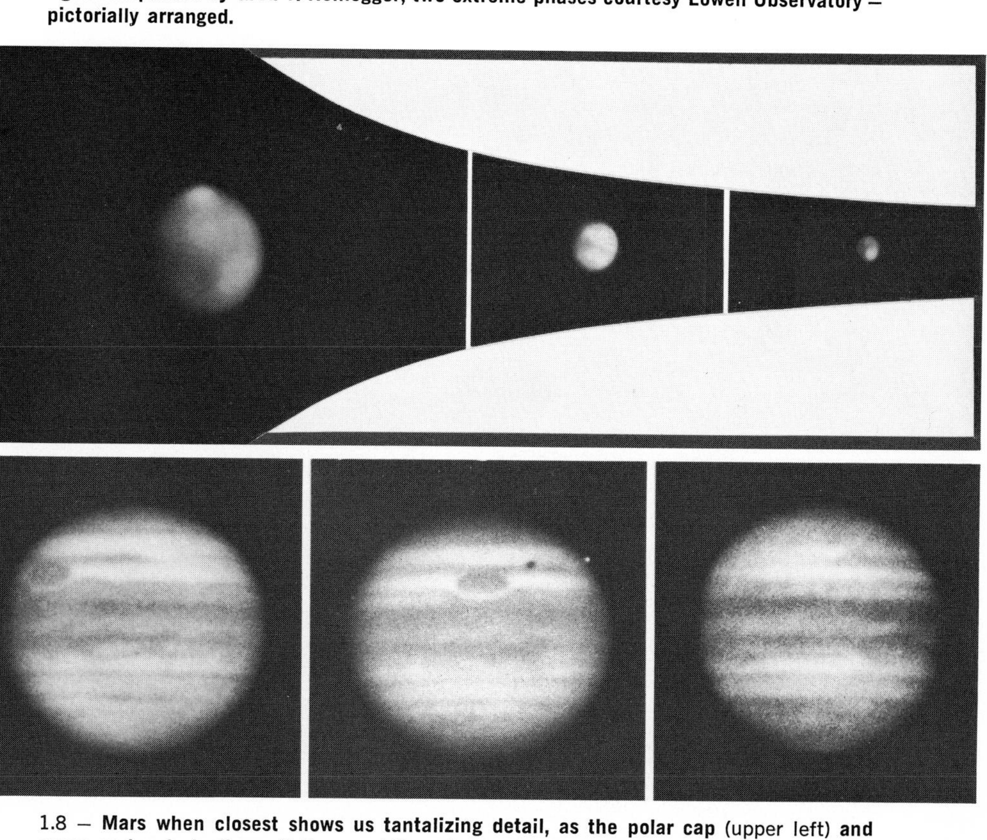

1.8 – **Mars when closest shows us tantalizing detail, as the polar cap** (upper left) **and green and red shadings. It then rapidly swings away to fade to a sixth the size. Larger photos by Dr. A. Dounce.** (below) **Jupiter's surface rotation is clearly indicated by the shifting great red spot. The smaller Moon II and its shadow appear in the central photo by H. Dall. Outer photos by T. Pope. All taken only a few days apart. (See color plates.)**

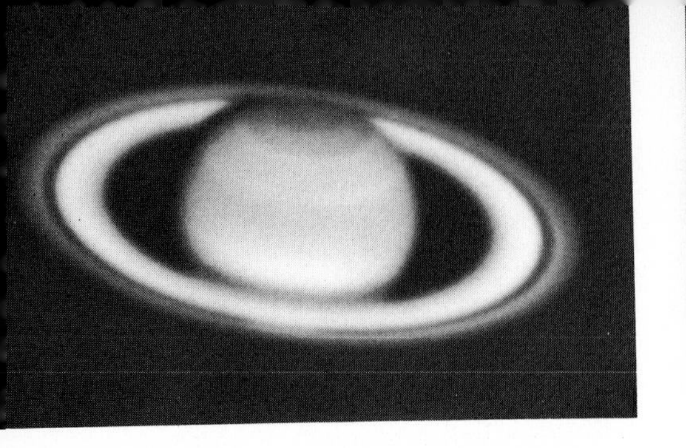

as the finest sight available to amateur telescopes and one that few veterans tire of seeing. The spectacular ring system and cloud banding change slowly as the planet tilts in its relation to us in its long thirty-year journey around the sun, as shown in Fig. 1.9. The change from near-maximum ring opening, as is splendidly shown at the left, to the minimum opening, when the plane of the rings starts to line up with our Earth (right), is nearly 7½ years. The latter condition, when for a time the rings are reduced to the appearance of a spike through the planet and totally disappear for several weeks, fell in 1966. These extreme phases *each* occur every 15 years or *twice* in Saturn's 30-year orbit. Probes will soon photograph directly.

Many amateurs find it most interesting to search telescopically for Uranus and Neptune as they wander among the stars and to note that they present discs of light to sharp lenses, not points of light as do the stars. Our nearest and farthest planets, Mercury and Pluto, yield little to most amateur scopes. Figure 1.10 (left) shows the tiny black dot of Mercury as it swings across the face of the sun and (right) the faintly recorded image of our remotest planet Pluto. Mercury is closely grilled under intense solar radiation, while Pluto, far out in frozen wasteland, receives only a pittance of the sun's light and warmth.

STARS THAT SHINE

The study of the stars and their constellations or grouping fills the history of astronomy. Names of constellations and signs of the Zodiac appear today in innumerable ways. Figure 1.11 illustrates the twelve signs of the Zodiac that we encounter so often, and stars and constellations inspire decorative themes on stamps and on many other objects to please our eyes.

1.9 – **Saturn slowly and majestically opens and closes its spectacular ringed system to our view as it moves ponderously in its 30-year orbit about our sun. These superb photos were taken at the Lowell Observatory many years apart to show the change.**

"It doesn't look any different through the telescope — maybe brighter." This often heard expression at first views of a single star is quite true, since stars are so distant they still appear as points of light even in the largest telescopes. However, strangely enough, what often appears as a single star to the unaided eye turns out to be a dazzling pair of twin stars. Some are true doubles or binary stars, swinging slowly about a common gravitational axis and sometimes even eclipse each other. Other are optical doubles, so called because, being nearly in the same line of sight, they appear close together, though one star is much farther away than the other. Double stars may have quite differing colors, as the spectacular Albireo, or β Cygni, shown in Fig. 1.12 (top right) exactly as seen in a 65-power telescope; their sparkling orange and blue images make a striking view. If we turn to the pair of stars Alcor and Mizar as they visually form the pair in the bend of the handle of "the big dipper," the 65-power telescope miraculously transforms Mizar into a sparkling double star. Certain double stars, having appropriate angular separations, make fine test objects against which to check the defining quality or resolving power of telescope lenses and mirrors. Many stars actually pulsate or vary in light intensity in strange and unusual ways of great interest to observers of variable stars.

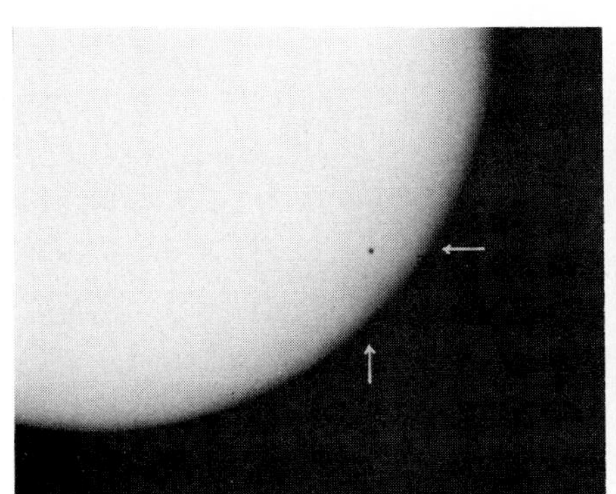

1.10 – (left) **The small black dot is our innermost planet, Mercury, photographed in transit across the face of the blazing sun by Paul W. Davis.** (right) **Arrows point to the dim image of distant Pluto in its cold outermost orbit. The bright star Delta Geminorum shows four diffraction spikes from equipment supports. Lowell Observatory photo.**

1.11 – **Twelve figures or signs of the Zodiac are constantly associated with our skies throughout astronomical history. These are strikingly shown in gold with white symbols on a dark blue one pound 1961 Israel stamp.** Counterclockwise **starting with 12 o'clock the signs are: Aries, Taurus, Gemini, Cancer, Leo, Virgo, Libra, Scorpius, Sagittarius, Capricornus, Aquarius and Pisces.**

It wouldn't mean much to give estimates of the enormous number of stars visible in the sky. Let's look instead at a star-rich portion of our sky in the form of our Milky Way, as it spreads like a phosphorescent ribbon of stars across the heavens in Fig. 1.12 (bottom). When we look more closely at only a small section of it in the region of Sagittarius, as shown

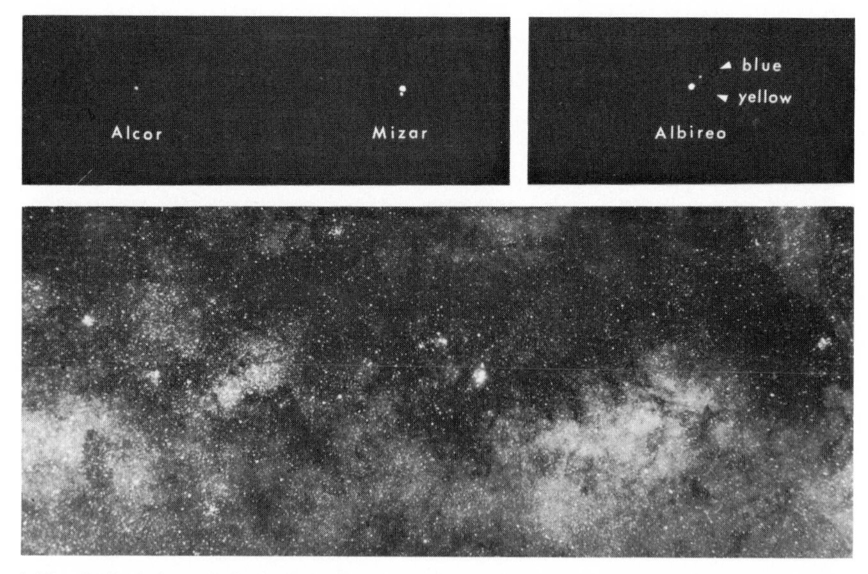

1.12 – (below) **A myriad of stars glow in our Southern Milky Way. Photo by the author.** (above) **Many stars are not single but colorful doubles, as is Mizar of the Alcor-Mizar pair in the bend of the dipper handle, and the flashing blue and yellow double star Albireo. From Anscochromes by Ralph Dakin.**

in Fig. 10.10, we will see glowing white clouds of glittering stars that literally light up the night sky. Yet the Milky Way is only a part of the galaxy of stars that contains our solar system of sun and planets. As we move out into space there are innumerable other galaxies just as immense but dimmed by their vast distance. Truly, the expression "countless stars in endless space" is a most appropriate summation.

Besides the configuration and areas of stars known as constellations, often best observed with the naked eyes, there are groupings of stars that always make interesting observing for both beginners and experienced astronomers through the smallest of optical instruments. The well-known Pleiades or "seven sisters" and the double clusters of Perseus, each a jewel-like mass of stars, are shown in Fig. 1.13. These are fine views in ordinary binoculars and striking objects in low-powered telescopes. We can also view loose, open clusters of scattered dazzling stars and compare these with a tightly packed glowing mass of thousands of stars as seen in the well-known great globular cluster M13 in Hercules.

NEBULAE AND GALAXIES

There are many strange glowing masses of gaseous material that take intriguing, mysterious shapes in the black night skies. These nebulae may often be the glowing remains of a cataclysmic stellar explosion that occurred ages ago and is still expanding into space. But all nebulae are not bright; sometimes dark obscuring masses of dust hang like black curtains in the sky to cut off light from the background stars. Such nebulae are clearly depicted on the stamps heading this section. In Fig. 1.14 we see well-known bright and dark nebulae available visibly and photographically to amateurs. The Orion nebula appears to contain obscuring clouds among its outstretched fingers of glowing gas. The great ring nebula in Lyra has an interesting central star. Most likely an explosion of this long ago started the outrushing, bright gaseous ring so striking in a dark sky. A black nebula in Orion rears up its horsehead shape of obscuring dust to cut off the light from its backdrop of stars and glowing gas. "Cold" cameras extend our grasp of these masses.

1.13 – **Dazzling star configurations, and both loose and tightly clustered masses of stars, dot the skies. In these photos we see the sparkling Pleiades group, by S. R. B. Cooke; the great Double Cluster of Perseus, by W. H. Shields; a loose open Cluster HVI30 in Cassiopeiae, by E. C. Silva; and the great condensed globular cluster Hercules, by Evered Kreimer.**

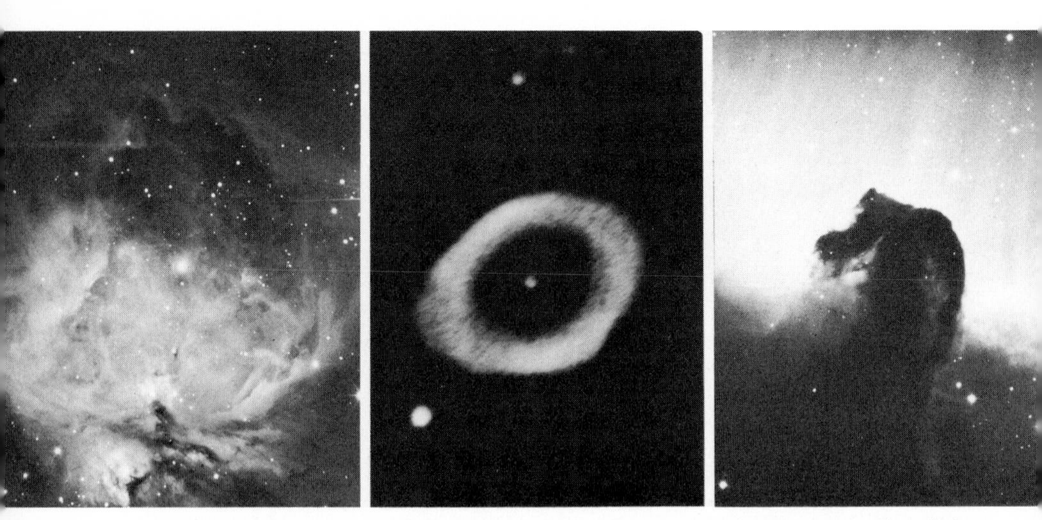

1.14 – **The bright gaseous Orion Nebula thrusts out its gleaming fingers in the photo** (left) **by Evered Kreimer, and the glowing ring nebula in Lyra** (center) **appears as a giant smoke ring, by C. P. Custer, M.D. Black obscuring masses of dust rise up in front of a glowing, star-filled background to shape the famous Horsehead Nebula, by the 200-inch at Mt. Palomar, photo from the Mt. Wilson and Mt. Palomar Observatories.**

While some nebulae present fine views through the amateur's telescope, others only appear as dim patches of light. Here the light-accumulating power of the "cold" camera presents these nebulae in their true grandeur, as may be seen in photos in the chapter on photography. (See color section.)

Many of the glowing objects in the skies termed nebulae by the early astronomers turned out to be galaxies of spiraling stars when checked with powerful new telescopes and cameras. These "island universes" far out in space often bear marked similarity to our own galaxy, the Milky Way system. Very well known to all amateur observers is the great Andromeda galaxy shown in Fig. 1.15, whose swirling arms have been resolved into individual stars. The Whirlpool nebulae shown is a small but bright object, yielding its spiral structure readily to the camera. Some galaxies photograph as spinning pinwheels, as shown in the center stamp above, and others take strange lenticular shapes, or hang nearly edge on as spikes of light in the sky. Other examples may be seen in Chapter 10.

CELESTIAL PHENOMENA

Many visitors to our night skies come unheralded. A meteor burns through our atmosphere, often exploding like a bomb; auroras quietly glow and rustle like fluorescent silk curtains in the sky; and silent comets swing slowly and majestically through the night, growing and receding in their path around the sun. All are spectacular and provide the artist with fine themes. Many occur in our own rarified upper atmosphere and can be seen with no optical aids or a pair of ordinary binoculars.

In Fig. 1.16 we see a Perseid meteor dashing down the Milky Way, disintegrating in a few short bursts of light. We also see two outstanding examples of the unusual and differing shapes streaming comets can take as the sun excites their thin atmospheres of molecules to glow as gossamer filaments of light attached to a gleaming head. A solar-inspired terrestrial aurora borealis with its soft pastel green and bright red arcs and draped curtains hanging down to the horizon can often put colorful firework dis-

1.15 — **Masses of stars spiral out into the glowing arms of these gigantic galaxies.** (left) **The great Andromeda galaxy is another universe of stars floating out far beyond our own Milky Way galaxy. Photograph by Alan McClure.** (right) **The bright Whirlpool Nebula in Canes Venatici clearly shows its attached smaller nebula. Photograph by Evered Kreimer.**

plays to shame. Black and white photos do not do justice to the delicately colored auroras, which are beautiful to watch and perfect subjects for modern ultra-speed miniature lenses and fast color films. In auroral displays, great starshell-like bursts of colored light may surge directly overhead to brighten, spread out and slowly flicker away — to be suddenly repeated in a cycle of new splendor. (See color section.)

Many other celestial phenomena occur. In Fig. 1.17 the shadow cone extending down to the earth at solar eclipse totality is clearly shown. "Sun dogs," zodiacal light and multicolored halos may appear. Fig. 1.18 is a copy from a vivid color photo of a tremendous ringed halo about a cloud-hidden sun near Stockholm, the sight of which, with its deep purplish blues and darkened azure inner sky, far exceeded even the color film's record. Note the sun's small disc near the bottom center. Two faint arcs shooting off at tangents from the top could be seen in the original transparency.

While some of the astronomical wonders presented in this chapter can be seen with the naked eye, a pair of binoculars or a modest telescope can increase your enjoyment of the night sky a hundredfold. Here is the world's most majestic open air theater — free of charge. Why not take advantage of it and explore outer space from your own back yard? Why not photograph these objects too? The corresponding sections of Chapter 10 devoted to astronomical photography supply further technical data on many of the photos presented here.

1.16 — **High speed and slow speed celestial events appear here.** (let) **A Perseid meteor streaks down the Southern Milky Way, as photographed by the author.** (center) **The Comet Arend Roland throws out a strange spiked nose.** (right) **Comet Ikeya swings through the sky with its gossamer tail streaming out behind. Photos by Alan McClure.**

1.17 — **The dark shadow cone formed at eclipse totality rushing rapidly along its path may be clearly seen as photographed from a plane by E. F. Carr, courtesy the Boston Globe.**

SELECTED OBJECTS FOR TELESCOPE VIEWING

After satisfying yourself with such planetary or lunar views as available, take a look through binoculars or *lowest* scope powers, at the Pleiades star group in Taurus, the Trifid and Lagoon gaseous nebulae (M20 and M8) in Sagittarius, and the Orion region. Now turn to the list below, an even dozen fine objects that can constitute "firsts" on anybody's observing list. They are presented as follows: a common designation and Messier (M) number; constellation; increasing right ascension (RA) in hours and minutes; declination (Dec.) in degrees above (+) or below (−) the celestial equator; general comments.

TABLE 1

Designation	Constellation	R.A. h m	Dec.°	Comments
Andromeda Neb., M31	Andromeda	0 40	+41.0	Great spiral galaxy of Andromeda*
Perseus Clusters	Perseus	2 18	+56.9	Grand Perseus Double Cluster*
Orion Nebula, M42	Orion	5 33	−05.4	Great galactic gaseous nebula
Sirius	Canus Major	6 43	−16.6	Brightest star, dazzling
Galaxies, M82 & 81	Ursa Major	9 52	+70.0	Distant spiral & edge-on galaxies**
Mizar (and Alcor)	Ursa Major	13 22	+55.2	Fine "easy" double, with nearby star
Hercules Clust., M13	Hercules	16 40	+36.6	Great Hercules globular cluster
Double-double	Lyra	18 43	+39.6	Two double stars in one field†
Open Cluster M11	Scutum	18 48	−06.3	Striking "open" cluster
Ring Nebula M57	Lyra	18 52	+33.0	Great Ring Nebula of Lyra**
Albireo (β Cygnus)	Cygnus	19 29	+27.8	Double, beautiful orange and blue
Planetary Neb., M27	Vulpecula	19 57	+22.6	Famous Dumbbell nebula**

*use low powers; **scopes 6-inch or larger; †high power

1.18 – **Great halos may encircle either our sun or moon. Here we see a bright-rayed halo about a partially obscured sun in a dark azure blue-black sky. The sun appears as a small white dot near the bottom center. From Kodachrome II wide-angle photo by the author.**

USEFUL AND INTERESTING INFORMATION SOURCES

Norton's Star Atlas, by Norton and Inglis (Sky Pub. Corp., Cambridge, Mass.) or H.A. Luft is a "must" for amateurs. The monthly maps and information in the two journals *Sky and Telescope* and *Review of Popular Astronomy* provide current guides.

Skalnate Pleso Atlas of the Heavens, by A. Becvar and Associates. More detailed than Norton's for the advanced worker. An outstanding guide to the sky, 21½ x 15½ inches, bound (Sky Publishing Corporation).

Two *elegant* large pictorial volumes are: *The Picture History of Astronomy* by Moore (Grosset & Dunlap), and *Astronomy* by Hoyle (Doubleday).

Stamp photos presented are by the author from his collection. Stamp collectors interested in pictorial stamps should refer to "Astronomy on Stamps" by A. P. Mayernik, 50¢ from Sky Pub. Corp., for full details on those presented here and many others. More and more stamps are appearing with "space" or astronomical subjects.

The amateur will be interested in *The Amateur Astronomer's Handbook* by James Muirden (Crowell).

Two recent very large and beautifully reproduced volumes are *The Atlas of the Universe* by Patrick Moore (Rand McNally) and *Astronomy* by Donald H. Menzel (Random House).

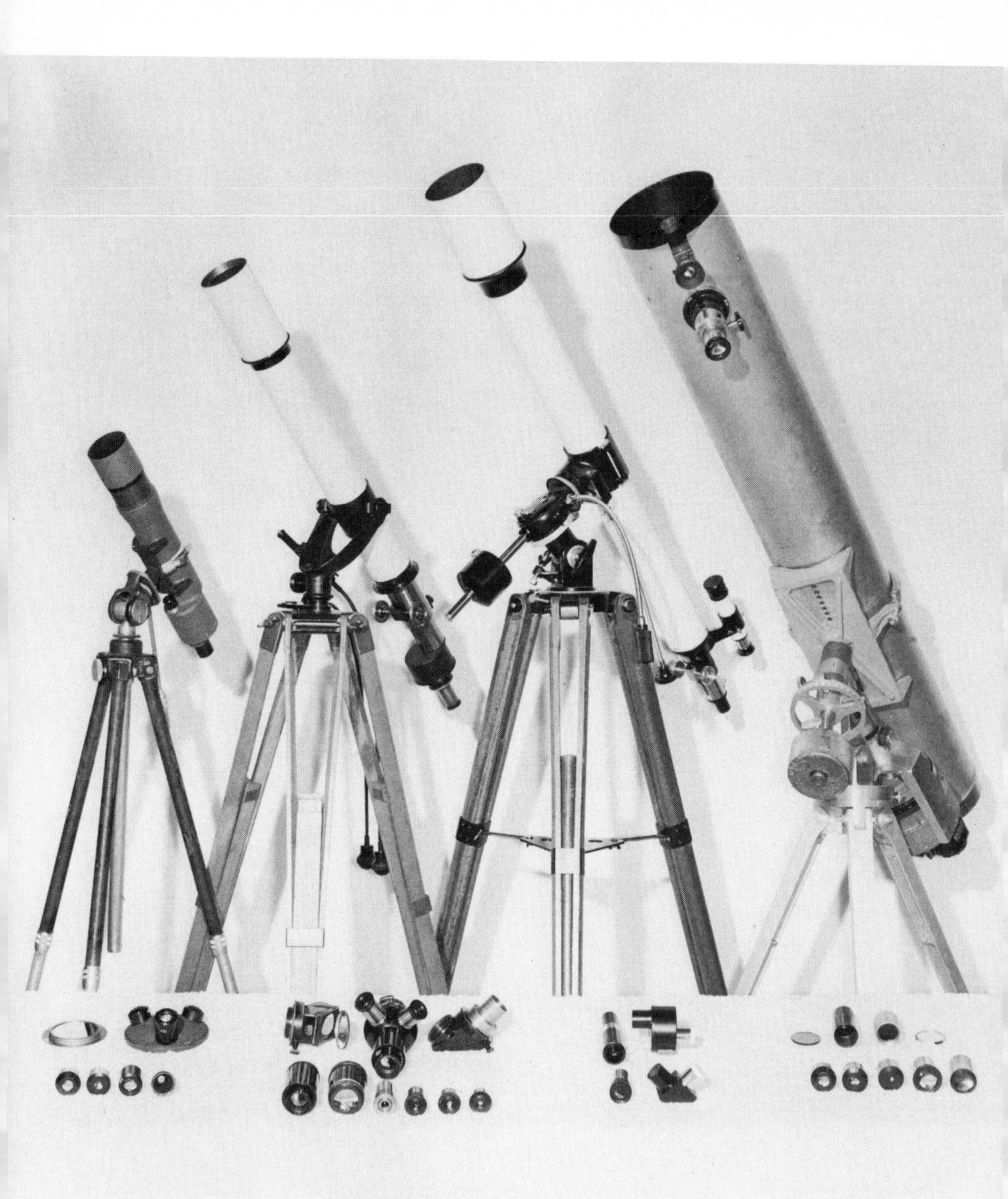

2.1 – **The four most practical types of telescopes for amateur astronomers.** Left to right: **the 60 mm. prismatic all-purpose scope, the 3-inch refracting lens yoke mounted altazimuth terrestrial–celestial type, the 3- or 4-inch refracting lens equatorial mounted telescope and the 4- or 6-inch reflecting mirror equatorial telescope. Not recommended, are 3-inch reflectors as they have excessive diagonal obstruction. Each has useful accessories, as shown. Photo by the author.**

2

TELESCOPES — GENERAL

Before we can see most of the wonders in the night sky, we must have the help of a powerful tool, the astronomical telescope, which effectively enlarges and brightens these objects enough for a closer view. When asked for a definition of a telescope, one prompt classroom answer was "a long tube with a lens at one end and an astronomer at the other." This is about as good a starting definition as any, since many of us think of lenses when telescopes are mentioned, even though mirrors can be substituted for lenses in both amateur and professional instruments.

Let's look at some representative examples of precision small and medium-sized telescopes in Fig. 2.1. The three at the left have refracting lenses at the front, and are commonly called *refractors*. The larger instrument on the right has a concave mirror at the bottom to replace a lens as the light collector (at this point you may wish to glance briefly at the first figure of the next chapter). Such mirror instruments, called *reflectors,* often represent a best investment for advanced amateur astronomers. Many astronomers assemble telescopes from component parts, and some even grind and polish their own mirrors and build their own tubes and mountings.

In all telescopes the lens or mirror functions as a receiver of the light from a distant object, but contrary to popular belief this primary lens does not magnify the object. Rather it collects and bends, or refracts, the light to form a relatively small image at its focus (near the rear of the tube of the common refracting telescope, or near the top of the reflecting type). Here a compact high-powered lens or lens set, called an eyepiece, applies the all-important magnification to the image.

How much more we can actually see of the object we are studying through a telescope depends primarily on the size of the main lens, or mirror. Hence the long-established custom of designating astronomical instruments by the size (diameter) of this primary optical element, usually in inches but sometimes in millimeters (mm.) in smaller ones (25 millimeters to an inch). Before we become too fascinated with this "the bigger the better" line of thought, let's hasten to add that much depends on the perfection of the lens or mirror and also on the highly variable terrestrial atmospheric conditions that quite often severely limit the size of telescopes for practical observing. The instruments illustrated ranging from about 2½ in. (60 mm.) to 6 in. in aperture (lens or mirror diameter) are very practical for amateur astronomers. More on sizes and types later.

Because of the high magnification required for celestial viewing, astronomical telescopes cannot be successfully hand held and require at least a firm tripod support. A firmly mounted instrument assures a sharp image and frees the hands for swinging the tube about, using the finder, focusing the eyepiece, adjusting circles or drive, etc.

MOUNTING TYPES

There are two common types of heads or mountings to hold the telescope tube firmly to the tripod or pier, yet permit it to be moved about smoothly to point at any portion of the sky. The first type functions exactly like a camera pan-tilt head and has two degrees of motion, one vertically (in *altitude*) and the other horizontally (in *azimuth*) — hence the name *altazimuth* applied to it. In fact, a camera pan-head and tripod make fine small telescope mountings, as shown at the left of the figure. More refined altazimuth mounts are most often of the yoke type, as shown second from the left. These are quite satisfactory for small instruments of the refracting lens type often used with terrestrial eyepiece units for looking around the countryside, yet they can still be turned to the skies at night by amateur observers. All goes fine with this type of mounting until we increase the instrument's magnification to powers of 150 or more. At this point our earth's steady rotation on its axis becomes startlingly apparent. The object now no longer appears stationary but marches rapidly across the field of view, requiring constant and annoying readjustment of the instrument. Something more is required of mountings for high-powered telescopes than an altazimuth provides.

Let's take a closer look at the cause of our problem. As our earth turns, its axis of rotation points steadily to an area in the sky called the celestial pole quite near to Polaris, the so-called North or Pole Star. This star is shown as it appears in our northern sky at the left in Fig. 2.2, where it may be readily located (upper center) by its being in line with the two stars forming the outer edge of the well-known "big dipper" in Ursa Major, as well as by its relation to the W-like configuration in Cassiopeia almost opposite to it. Amateur astronomers and astrophotographers soon learn that Polaris is actually not quite at the true pole, but about a degree from it

2.2 – The pole star, Polaris, is actually almost one degree from the true pole, which it circles, as shown in the five-hour exposure above. The relation of Polaris to the star configuration of the Big Dipper and the W star formation of Cassiopeia are clearly depicted on the stamp, a blue-gray commemorative stamp to the Tokyo Astronomical Observatory. Circumpolar star trails by Robert Cox – note the meteor trail captured in the lower right corner.

at the present time. This is clearly demonstrated in the star trail photo presented above in the figure, where the star trails recorded by a fixed camera form arcs centering on the true celestial pole as the earth rotates. The brightest small curved arc near the center is the apparent path of Polaris swinging about the true celestial pole during the five-hour exposure.

To nullify the effect of the earth's rotation most serious amateur astronomers turn from the altazimuth mounting to one called an *equatorial* mounting. It also has two degrees of motion with axes at right angles to each other, but one axis maintains a fixed special position while rotating, that of constantly pointing towards our celestial pole. The basis for this type of mount is shown in the Russell Porter sketch of Fig. 2.3. Here we see the *polar axis* about which the tube can be swung in *right ascension* is parallel to the earth's axis; both in turn point toward the celestial pole in the sky. The angle A between the telescope's polar axis and the northern horizon is always adjusted to be exactly equal to the local latitude, as shown by the other angle A of the diagram. Now if one slowly turns this polar axis at the rate of one revolution per day in the reverse direction to the earth's rotation, no matter where the telescope is pointed in the sky by tilting on the second, or *declination* axis, it will steadily follow an object to make it appear

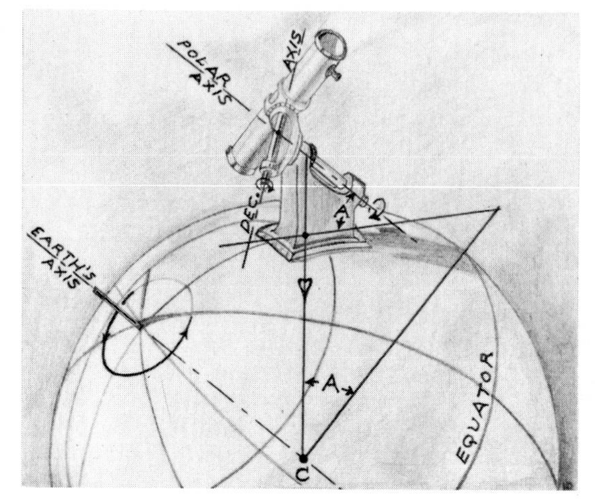

2.3 — **Russell Porter informatively illustrates the how and why of the equatorial type mounting in which the polar axis always points to our celestial pole. When this axis is turned opposite to the earth's rotation at one revolution per day the telescope remains pointed at celestial objects. See text. Adapted from Amateur Telescope Making Book One, courtesy Scientific American Publishing Co.**

perfectly motionless or fixed in the field of view. Now the fact that a scope's polar axis and the earth's axis are several thousand miles apart is quite insignificant in observing objects millions of times farther away. Precise polar axis adjustment is even more of a "must" for astrophotographers. The refractor and reflector at the right of Fig. 2.1 are on equatorial mountings.

TELESCOPE SIZES

Since the *size* of an astronomical telescope is so closely associated with highly important characteristics, such as its performance, convenience of operation, portability and cost, for the purpose of this book all instruments will be divided into three arbitrary groups: *small, medium* and *large* telescopes. The dividing lines based on the above factors are not sharp but do exist for most who know the field, though there is no exact agreement. Each size group serves differing needs, which can range from a youngster taking his first look at the moon through a small "spy glass" to a large high school wanting its own observatory with a 20-inch reflecting telescope.

Since the actual physical diameter (aperture) of the primary lens or mirror is most closely tied to a telescope's performance in resolving fine detail on celestial objects viewed (in fact limits the instrument's *useful power* or magnification), it is necessarily an important unit of measure. Yet one must balance this against such other factors as the overall size or bulk and the type of telescope we are considering. Refracting lens telescopes have much longer tubes, taller mountings, and cost much more than do reflecting mirror telescopes with the same effective aperture. Each type has many advantages — and some disadvantages. The classification below considers *all* the factors concerned.

Small Telescopes comprise refractors from the smallest one-inch diameter lenses to those of about 2½-in. (60 mm.) size. These are often on altazimuth mounts or fitted with an adapter or threaded socket for use with a camera pan-head and tripod. These are often used both astronomically and terrestrially (with an erecting eyepiece system). They are exceptionally portable and usually quite sturdy and trouble free. Secondary "finder" telescopes are not required on lower-powered models.

Small reflectors range from about 4-in. to 6-in. in mirror diameter. They should rightfully have astronomical type equatorial mountings (even if simple), since they have limited value for terrestrial use. These small instruments often serve well as a youngster's first telescope, particularly the refractors.

Medium Sized Telescopes cover refractors of 3-in. to 5-in. aperture. These are still quite portable with tripod supports. Such instruments, if well made, fully justify an equatorial head, even one with an electric drive. They are easy to use, stay in adjustment, and with a prism erecting eyepiece can still be used terrestrially. Such instruments have marked advantages in educational fields.

Reflectors of 6-in. to 10-in. mirror sizes on firm, clock-driven equatorial mounts make up a most popular group of medium-sized instruments specifically intended for astronomical use. They offer the most for the money in true astronomical "seeing power" for amateur astronomers. Most amateur telescope makers build medium-sized instruments, and justly so, since they represent a best investment in both time and money. I like to encourage the growing group that neither totally build a telescope (including mirror), nor buy a strictly commercial instrument — a group that assemble a telescope from any available part source and learn a lot in doing so. The reflector at the right of

2.4 – A very practical-sized, portable, equatorially mounted 4-inch short-focus refractor with synchronous clock drive, partially built and assembled from a wide range of miscellaneous ancient mechanical parts, surplus optics, etc., by Ralph Dakin and the author.

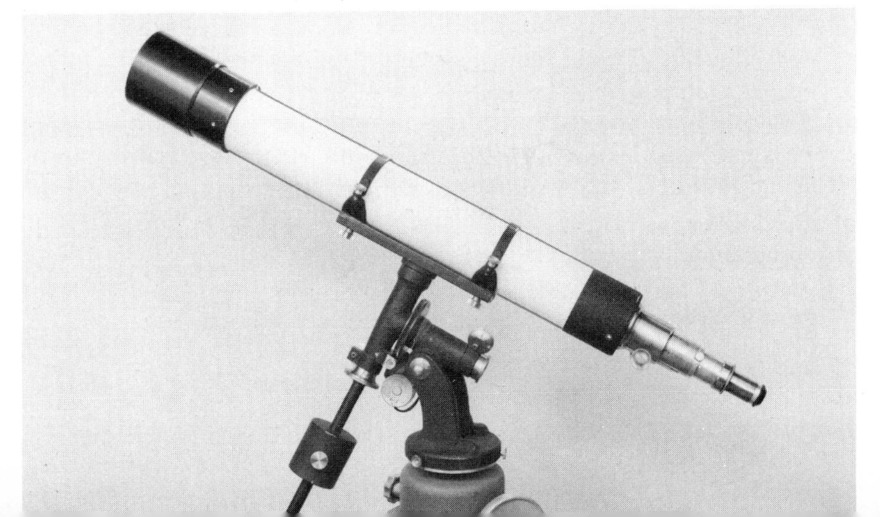

2.5 — A neat roll-off-roof observatory built by Paul W. Davis houses a Cave equatorial 12½-in. reflector. Such a unit is ideal for an individual, a club or a small school. Note the fine display of astrophotographs taken by Mr. Davis.

Fig. 2.1 was assembled from parts, though I also ground and polished the mirror. The fine portable short-focus 4-in. refractor shown in Fig. 2.4 was partially built but mostly assembled by Ralph Dakin and myself from a multitude of miscellaneous parts, some of quite ancient origin. It's a pleasure to use.

Larger Instruments for serious amateur and teaching use cover 6-in. and occasionally 8-in. refractors of normal focal lengths. These deserve a solid fixed equatorial mounting and reasonable protection, if not an observatory. The 6-in. aperture $f/10$ or "short focus" type can still be portable. Such refractors can be quite expensive unless you assemble these from available components, as you readily can the 6-in. size.

2.6 – **Over two centuries of progress in the development of telescopes is strikingly illustrated here as we turn from the 1737 Gregorian astronomical–terrestrial metal mirrored telescope in brass by James Short, to the modern compact, precision lens–mirror Questar telescope designed by Lawrence Braymer. Photo by the author.**

Reflectors of 12-in. and up are suitable for serious amateurs, large amateur groups, and large high schools, or small colleges. The fine, neatly housed 12-in. Cave reflector shown in Fig. 2.5 serves very well. Larger sizes are certainly not needed for general observing. All deserve to be suitably housed, kept in adjustment, and periodically inspected. Too often individuals allow fine large reflectors to become "white elephants"—just too much trouble to put into operation. For these an optically superb 8-in. conveniently housed with many accessories might give more pure pleasure. Still larger instruments really require devoted individual attention through all phases from acquirement through housing, operation and adjustment.

Truthfully, a medium-sized reflecting or refracting telescope, *if of high quality* and on a solid equatorial mount electrically driven to follow the stars, is a most powerful and efficient tool to observe and study (or photograph) endless fascinating astronomical objects over many pleasurable evenings. Telescope designs have now been modified to produce compact portable forms using both lenses and mirrors in combination. I took the photograph shown in Fig. 2.6 to illustrate two centuries of progress in compact telescope design — from the first practical reflecting astronomical–terrestrial telescope by James Short in 1737 to an ultra-modern precision lens–mirror telescope, the Questar, the "Rolls-Royce" of scopes. Such lens–mirror telescopes can perform well, but only if built to precise specifications and close tolerances, either by the manufacturer or the amateur.

Astronomical societies' "observing" evenings, or just friendly get-togethers add interest and companionship to what could be a lonely hobby. Such gatherings provide just the opportunity to demonstrate the merits of a newly acquired telescope or gadget, and to point out a new comet to a veteran observer or a dazzling star group to the first nighter. In Fig. 2.7 we see a few members of the Rochester Academy of Science group gathered about George Keene's earlier yoke-mounted 12-in. $f/4.3$ visual-photo reflector for a pleasant evening of observing, with an appropriate "coffee break—discussion period" of course. Mr. Keene is not only a builder and astrophotographer, but is widely known as the author of the book *Stargazing with Telescope and Camera.* He has now completed a 20-in. scope.

At the end of this book will be found a suppliers name list, and comments and notes on journals and books.

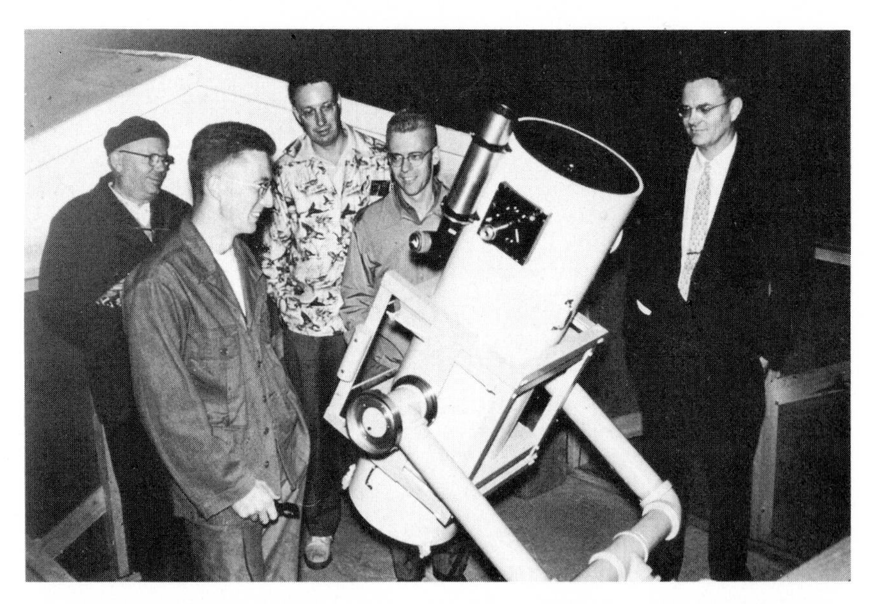

2.7 – A backyard observatory can serve as a pleasant meeting ground for amateur astronomers. Here George Keene puts his home-built 12½-in. reflector into operation for other members of the Rochester Academy of Science. A converted silo now houses Mr. Keene's 20-in. scope. See Sky and Telescope **Vol. 50, no. 1, July 1975 for new scope, and color section of this book for fine sky view from inside.** Left to right: **Paul W. Davis, George Keene, Ralph Dakin, John Schlauch, and Dr. Alexander Dounce. From Kodachrome II by the author.**

3

LENSES VERSUS MIRRORS

HOW TELESCOPES WORK

The inside story behind the way astronomical telescopes function is amazingly simple, and you'll enjoy acquiring one much more if you take a few minutes to learn how they work and see how they grew from a small beginning. If at night you take a lens as shown at (L) at the top of Fig. 3.1 (a) and hold it up toward the moon, it will collect and bend the light to form a bright image of the moon (upside down and reversed) on a piece of tissue paper or ground glass at the lens' focal point (I). If we mount such a refracting lens, called an objective, at the front of a tube at (O), as shown in sketch (b), and take a small high-power magnifying lens unit (the eyepiece at E) and focus it on the image (I), we will see at the bottom end of the tube a magnified view of the moon. We have just assembled an astronomical telescope called the *refractor*. That the image is inverted is of little concern to astronomers, since the terms "up" and "down" have little meaning in the skies, as long as one has proper reference points.

At (c) in our figure we have a totally different astronomical telescope that uses an appropriately shaped curved mirror at the bottom of a tube instead of a lens at the top. This concave (parabolic) mirror at (M) collects the light just as a lens does, but here the similarity ends, the mirror reflecting the light back up the tube to a focus at (I_1). Since this image cannot usually be examined at this position in the tube without one's head obstructing the incoming light, a small flat mirror or diagonal is placed at (D) to reflect the light beam at right angles out through a hole in the side of the tube to (I_2), where it may now be easily examined and magnified by the eyepiece (E). This instrument is the widely used *Newtonian reflector,* named after the Englishman who developed this form.

Compactness in larger instruments, or portability in smaller ones, is gained by the optical design shown in sketch (d) of the figure. Here again a parabolic mirror at (P), now termed the primary mirror, collects the light to send it up the tube toward a focus. However, before it reaches the focal point a small, round, convex secondary mirror at (S) catches the light and reflects it back down the tube as a much slimmer cone of light rays to squeeze through a hole cut in the primary mirror (P) to reach the eyepiece and the eye. This telescope, the *Cassegrainian reflector,* is named after its French inventor.

Users of refracting or reflecting telescopes with eyepieces at the rear often insert a small flat mirror unit (called a star diagonal) just ahead of the image to reflect the light up at right angles, as shown added to the Cassegrainian reflector (d), which places the eyepiece and the eye in the new position indicated, and avoids much "neck wringing" caused by looking into the eyepiece at an uncomfortable angle. Such an added star diagonal erects the image top to bottom (but not right to left; neither do twin-lens reflex or non-pentaprism reflex cameras using similar diagonals).

Recently medium-sized telescopes similar in appearance to the Cassegrainian have been developed in which a thick curved miniscus lens has been added, as indicated lightly sketched around (S) in the drawing (d), to make fine, compact, closed-tube lens–mirror systems — the lens serving to replace interfering metal secondary mirror supports. Or a central area of one of this lens' own surfaces has a reflecting coating to serve the function of the secondary (S). Since the mirror surface curves of these can differ slightly (but most significantly!) from those of the classical Cassegrainian parabolical primary and hyperbolical secondary mirrors, these lens–mirror catadioptric systems appear in many modifications under such names as Maksutov, Dall-Kirkham, Schmidt and others.

These are the broad basic systems of practically all the telescopes useful to the amateur astronomer, regardless of the multitude of variations in physical design, mounting types, sizes or prices.

EARLY ACCOMPLISHMENTS

Now that we've seen how modern telescopes work, it should be of interest to turn back for a brief glance at a highpoint or two in the history of their development. Your friends will expect you, as an amateur astronomer, to know some of the early accomplishments and will respect you for knowing them. The story of the development of telescopes touches on many nations and is filled with discoveries associated with great and colorful personalities. These discoveries often made long strides in the step-by-step scientific advancement to the present space age. Many beautiful, intriguing and skillfully engineered instruments of wood and brass still speak for these individuals and their achievements.

The trail begins with a Dutch lensmaker, Hans Lippershey, who in

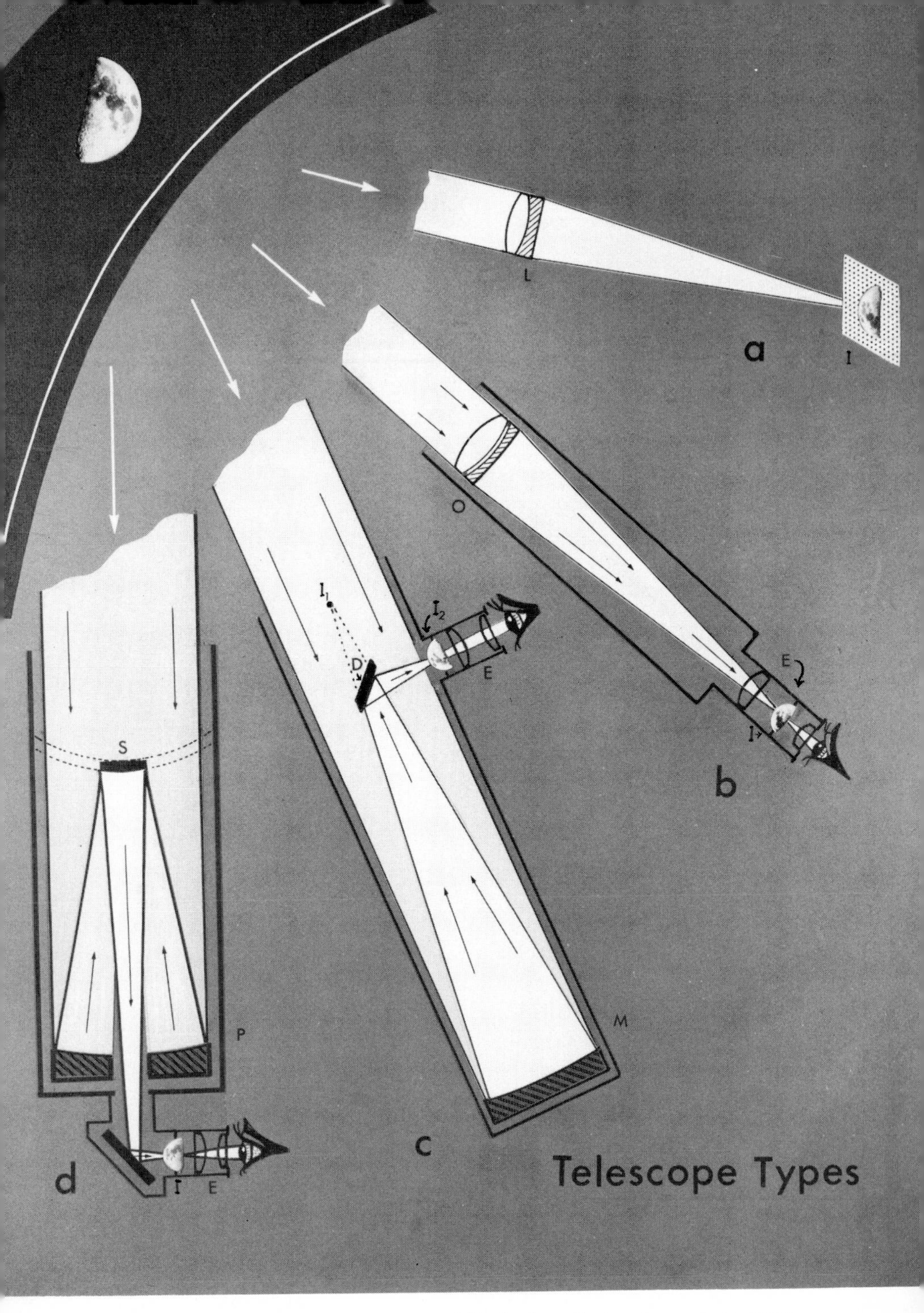

3.1 – Most astronomical telescopes use one of the basic optical systems shown above. Each differs considerably from the other, with many advantages and some disadvantages. See text. Drawings by the author.

3.2 — Galileo's slim-tubed original instruments with their thumb-sized lenses introduced the telescope to astronomy in 1609 — an important event in man's conquest of space. Galileo appears here on a violet and brown Italian commemorative stamp. Photo courtesy Instituto E Museo Di Storia Della Scienza, Florence, where these instruments may be seen. Cracked original lens is in the plaque.

about 1608 reputedly saw a magnified image of a church steeple when sighting through a concave spectacle lens placed near the eye and a convex one held at arm's length. From here we turn to Padua, Italy, and to Galileo Galilei's practical use of such simple lenses in slim tubes, as shown in Fig. 3.2, and then continue on a path touching many nations and individuals before finally arriving at our modern lens–mirror telescopes. This entire phase of history is condensed in a striking way by a single fine illustration (Fig. 3.3), which takes us from the early astronomer pictured with his long tube telescope right up to the pattern formed by light traveling back and forth in a modern compact telescope — one no larger than the book telling the early story, and one now used by Celestron, Criterion and others.

Following the trail of development we encounter a fascinating race for superiority between the refracting lens telescopes and reflecting mirror telescopes, one that swung back and forth over the centuries and actually still continues in amateurs' back yards. The very first forms of each type have become virtually extinct but are historically important. In the end, the true refracting and reflecting telescopes appear to have approached maximum sizes and worth in the 40-in. Alvan Clark lens in the Yerkes re-

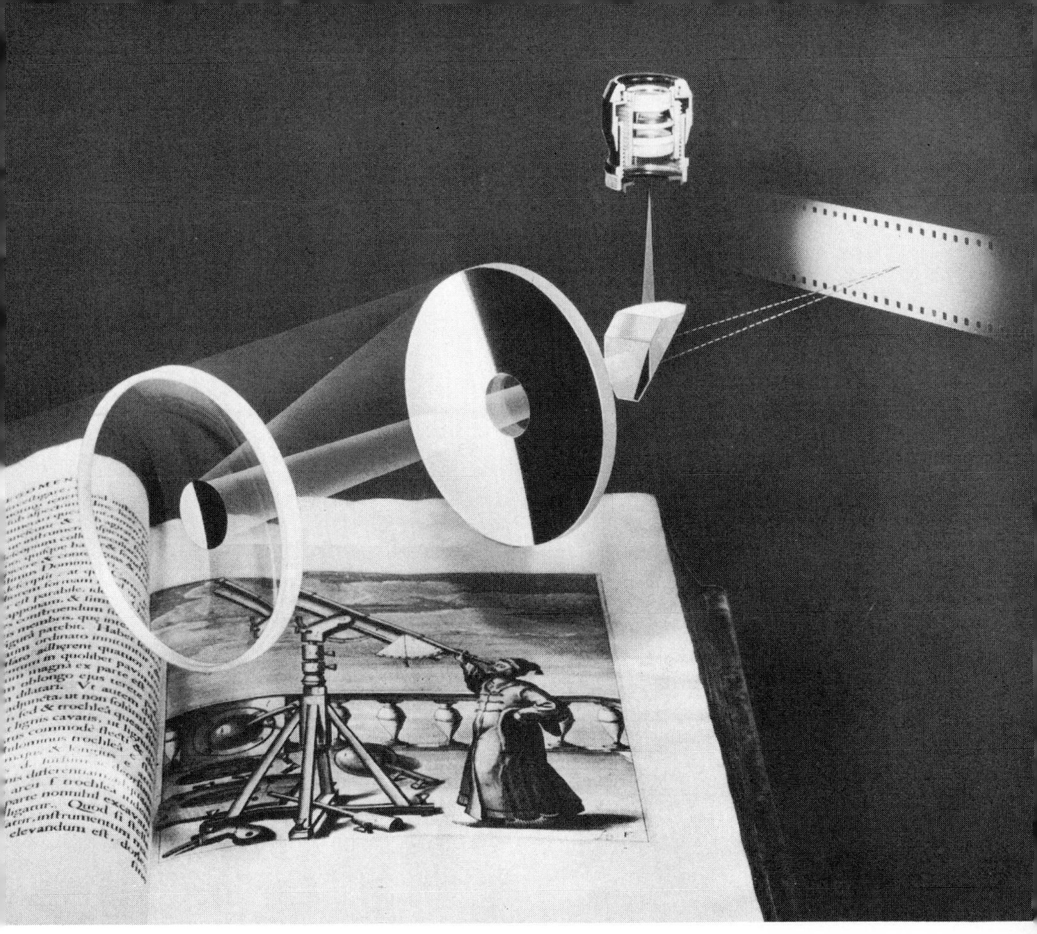

3.3 – This single photo tells the story of improvements in telescope optics over the past three centuries, from a single lens at the end of the long-tubed telescope of the early observers to modern compact lens–mirror optics – smaller than the ancient volume telling the early story. Courtesy The Questar Corporation.

fractor at Williams Bay, Wisconsin, and the 200-in. Hale reflector at Mt. Palomar, California — each a giant in its own right and each extremely useful. Strangely enough a wedding of the two forms now produces our modern superb lens–mirror telescopes.

Immediately following Galileo's introduction of the telescope to astronomy, many strove to improve the performance of the single lens objective. The best shape for it was soon well known, one that is quite convexly curved in front and almost flat at the back. This form gave somewhat better but not truly sharp images; furthermore, the images from such single lenses were surrounded by colored fringing. Figure 3.4(a) shows the reason. Light of differing colors does not come to the same focal point, blue light falling shorter than the red, producing the chromatic error that bedeviled single

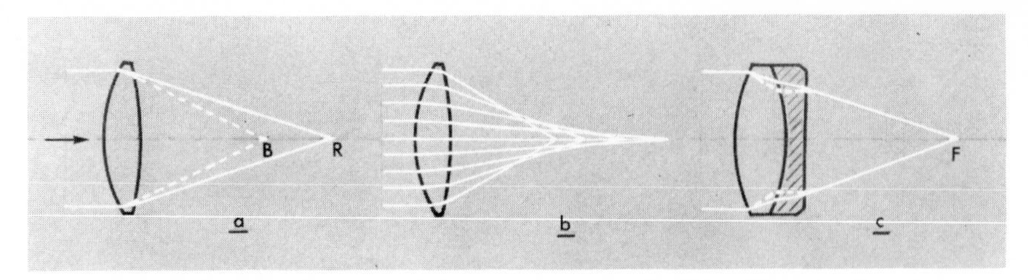

3.4 – **Lens errors illustrated.** (left) **Color or chromatic error – different colors do not focus the same.** (middle) **Spherical error – focus of edge light different from central light.** (right) **Combining crown and flint glass elements for a corrected "achromat." By the author.**

lenses. To make matters worse light of even a single color could not be sharply focused, because that passing through the outer edge of such spherically surfaced lens is bent more and reaches a focus sooner than light passing through nearer the center, as illustrated at (b). This error, called spherical aberration, prevented truly sharp images. It was learned that the situation could be improved if the focal length was increased with respect to the lens' diameter. This sparked a secondary race in which the avenue of longer and longer telescopes was explored to the very fullest by such men as Hevelius in Poland, Christiaan Huygens in Holland and Adrien Auzout in France. Single lens telescopes stretched out to become unbelievably long, awkward instruments swinging from poles hundreds of feet above the ground, as depicted in the rare engraving of Fig. 3.5. Believe it or not, such lenses performed quite useful tasks, as shown in the figure insert of an early drawing of the lunar mountains and Alpine Valley. It's quite interesting to compare this with the photo of Fig. 1.4. With such single lenses Huygens first made out the true form of Saturn's ring and discovered its brightest moon. The aberration of light was discovered, the diameter of Venus measured and a final touch added on the proof that the earth travels around the sun — all by James Bradley, an Englishman. These and many other discoveries are a credit to the perseverance of these pioneers and to their acute observations under most trying conditions, working with dim images from a tremendously distant aerially supported single lens. What a change from the tiny lenses and spindly tubes that introduced Galileo to the wonders of the heavens! Yet the earlier great astronomer Tycho Brahe, working to physical limits without lenses, would have given almost anything for even the simplest of these instruments. The long giant tubes are gone, but occasionally a smaller one may be found quietly resting, long forgotten, gathering dust in storage; see Fig. 3.6.

Before leaving the lensed instruments we must turn a moment to the other end of the telescope tube, the eyepiece end, where the astronomer

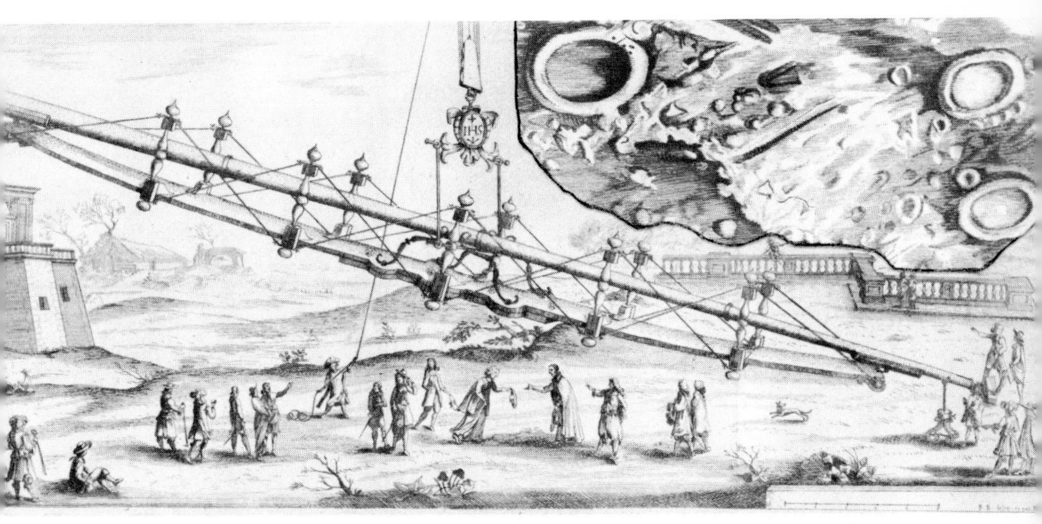

3.5 – **Fabulous lengths were reached in the single lens telescopes of early years, reputedly over twice that of this 75-foot monster. Yet real contributions were made: note Alpine Valley and lunar detail of early drawing** (inset). **From the ancient Italian volume** Hesperi et Phosphori nova Phoenomena, **by Blanchini, courtesy George Staack.**

3.6 – **A well-preserved ancient Venetian telescope with single lens in a slim, two-section twelve-foot wooden tube, quietly resting in dim storage as an artistic reminder of the extinct long-tubed telescopes. Seen courtesy of Gualtiero Schubert, Milano. From Kodachrome II by the author.**

or a camera must operate. The Galilean telescope used a single negative lens as an eyepiece, as shown at the left of Fig. 3.7. This with the eye produces a visually erect image (I), as indicated by the arrow in the diagram. This design was plagued by an extremely narrow field of view, about what one sees when trying to look through a long slim pipe. It's difficult to see how the early workers even located objects in the sky through long tubes swaying on poles. Johann Kepler, a German student of Tycho Brahe in Denmark, best known for his famous laws of planetary motion about our sun, improved things by suggesting the use of convex lenses, as shown in the central part of the drawing, which gave a considerably widened field but an inverted image, a factor of little concern to astronomers then or now. The small black arrow is the real image (I), and the larger dotted one is the apparent magnified image (M). In turn the Dutch mathematician and astronomer, Christiaan Huygens, better known for the clock's regulating pendulum and his discovery of the rings of Saturn and markings on Mars, again greatly improved eyepiece performance by using twin lenses arranged with the image between the lens, markedly widening the field of view and reducing eyepiece color error. Later Jesse Ramsden, an English manufacturer of quality instruments, reversed the front lens of the Huygens' eyepiece, changing it from a negative to a positive eyepiece, one where images or micrometer cross hairs fall conveniently in *front* of the eyepiece. Such Ramsdens are superior to Huygens' types for reflectors and are accordingly shown in Fig. 3.1(c) and (d).

In 1663 James Gregory, a Scottish astronomer and mathematician, designed the first reflecting telescope using mirrors, as shown at the right of Fig. 3.7. The concave parabolical mirror at the bottom of the tube reflects the light up to the focus, where the rays cross at a first focal point (f_1) for an inverted image, and then continue on to a small concave ellipsoidal mirror whose first focal point is also (f_1). This mirror in turn reflects the rays from its first focal point back down through a central hole in the main mirror to its second focal point (f_2) at the rear where the eye can examine a magnified image. Points f_1 and f_2 are the foci of the ellipse shown in the dashed line. Since this second image is erect, the Gregorian, like the Galilean, could be used terrestrially. Unfortunately, no one at this time had the skill to shape and match the mirrors properly, and finished but useless instruments became desk pieces. Beautiful mezzotints of the day show these small Gregorians being used as art props with famous personalities, about the only duty for which they were suited.

Meanwhile history amply records that in 1668 Sir Isaac Newton constructed the first practical reflecting telescope in the design shown in Fig. 3.1(c), in which a small flat mirror turns the light reflected forward from the main mirror out through the side of the tube, the well-established New-

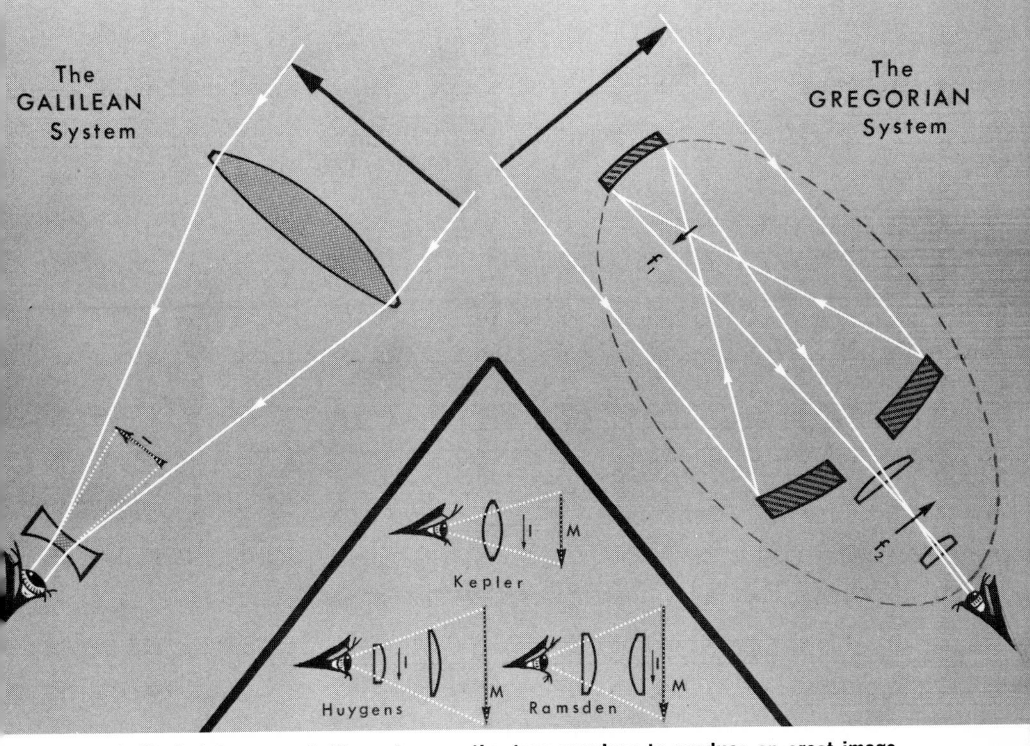

3.7 – **Galileo's telescopes** (left) **used a negative lens eyepiece to produce an erect image and could be used terrestrially. Gregory's reflectors** (right) **also produced an erect image for countryside viewing. Each system was a "first." Though now virtually extinct, they made great contributions to telescope development. Drawings by the author.**

tonian form most used even today for reflecting telescopes. Such reflectors have no color error. As with Galileo's first lenses, Newton's first polished metal mirrors were mere infants, approximately an inch in diameter, polished with putty powder (tin oxide) on hard pitch laps, which he introduced. One of Newton's first two original telescopes, as preserved by the Royal Society, is shown in Fig. 3.8, along with its inscription and a stamp issued in his honor.

We now move on to the ingenious Frenchman, Sieur Guillaume Cassegrain, a sculptor who in 1672 invented the shortened form of compound telescope shown in Fig. 3.1(d); however, Newton's sharp criticism of its faults, with no recognition of its merits, drove the Cassegrainian form into temporary obscurity. In 1721 John Hadley, the English precision craftsman and inventor of the sextant, following a three-year struggle, succeeded in polishing and parabolizing, possibly for the first time, a 6-in.-diameter concave mirror to a remarkable degree of perfection. In fact it was so perfect that when tested by the then esteemed James Bradley it stood right up in

3.8 — Sir Isaac Newton built the first practical reflector telescopes. (left) **The one preserved is quite small, only about 12 inches high in the position shown.** (right) **The inscription on its base. Mirror telescopes were to grow and grow, to culminate in the giant at Palomar. Newton is shown as he appears on a blue commemorative French stamp. Photos by C. Adair, permission of The Royal Society.**

performance in its easily handled 6-ft. tube against a 123-ft.-long refractor by Christiaan Huygens — a telescope 20 times longer and immeasurably clumsier. Here was real progress in telescope making, an event that possibly constitutes a first faint rumble of the battle of the larger telescopes soon to follow. Considering the date, it is startling to think that Hadley's reflector was almost identical in size and form to the most popular amateur observer's telescopes today, except that bright, aluminum-coated glass mirrors now replace the ancient speculum metal alloys of copper and tin that soon tarnished. But others could not acquire Hadley's skill, and it remained for another highly gifted Englishman, James Short, to revive the old Gregorian design in 1734, after it had lain dormant for over a half century. Short's deft fingers ground and polished the parabolic primary, and more importantly the small ellipsoidal secondary, to the high degree of perfection and matching required to produce for the first time a practical high-quality Gregorian telescope. He then developed a true equatorial mounting. James Short soon became famous for his many fine instruments, and in 1737 moved from Edinburgh to London to be elected a Fellow of the Royal Society, open an optical shop, and enjoy 30 years of prestige, even with the royalty. One of his fine instruments, engraved "James Short, Edinburgh 1737," was shown in Fig. 2.6. Since the image is erect in these instruments, they were used to view the countryside; in fact, they may be termed the first compact high-powered "spotting scopes." The one shown actually has "internal focusing" as do our latest and best small prism scopes, and there is even a scale at the front from which one can read the distance of the object in focus up to a mile away! And all this over two centuries ago. Both the Gregorian and Galilean types have now virtually disappeared from astronomical use, but not without contributions.

Our path turns back to the now neglected single lens refracting telescope, which had to be revived in some miraculous way before it could hope to compete again with the reflectors. Chester Moor Hall, an English gentleman and counselor, finally decided in about 1729 to ignore the long-held belief, arising from the mistaken views of Sir Isaac Newton, that lenses could not be improved — or perhaps he just didn't "know they couldn't be improved"! At any rate, Hall succeeded in combining crown and flint glasses, as shown in Fig. 3.4(c) to make the first color-corrected two-element achromatic lenses. This fortunate wedding of two differing kinds of glass had high potential indeed. Nevertheless, the poor glass available, lack of skill, and possibly inadequate publicity caused matters to rest until 1758 when John Dollond, a British optician, rediscovered, skillfully constructed, patented (after quite a legal contest) and exploited the first practical achromatic telescope lenses. With these new crown and flint glass "achromats" focal lengths could not only be remarkably reduced, but the resultant images were ultra sharp and almost color free. The day of the ridiculously long telescopes was over, and the refractor was now ready to vie again with the reflector for a place in the march toward bigger and better telescopes. In fact the refractor now had the distinct advantage of permanence over the reflector, with its delicate metal mirror surfaces. These required constant attention, since they rapidly darkened with tarnish, necessitating tedious repolishing and refiguring. During the middle years of telescope making a great many fine small instruments were turned out by manufacturers whose names were derived from their prominent founders, as Dollond, Ramsden, Short and others. Figure 3.9 shows two from this age still in fine condition.

Meanwhile, the mirror too was to receive a tremendous boost. Sir William Herschel, a musician, apparently as a hobby, decided to make a reflecting telescope because he could not afford the then enormously high-priced instruments. Herschel was an imaginative, exacting and tenacious worker. He was personally to give the mirror telescopes such a start in the race for size alone that the refractor never really caught up. After about two years of tedious effort he had acquired the art of precision mirror making and proudly displayed a couple of superb reflectors of about

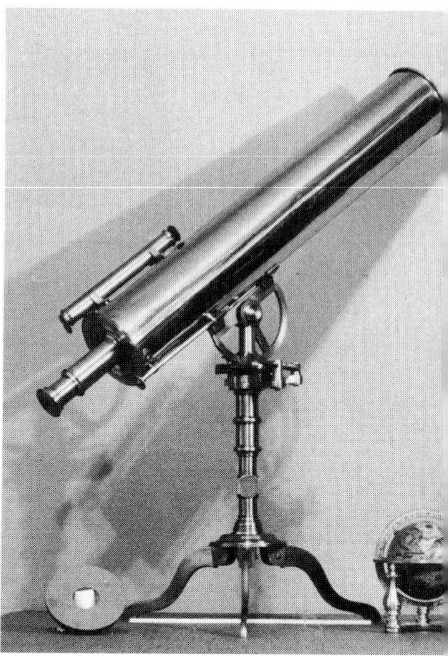

3.9 – Jesse Ramsden is widely known for his eyepiece design. (left) A 60 mm. refractor in brass from about 1775 signed by him, often on display at the Strausburgh Planetarium, Rochester, N.Y., as presented by the author. (right) An elegant polished brass Gregorian with metal mirrors in original condition and in perfect working order. Made in 1804 by B. E. Van der Bilt of the famous Dutch instrument-making family. Photos by the author.

3.10 – This 6-in., 7-ft.-focus reflector by Sir William Herschel about 1778 and an earlier similar one by John Hadley, both Englishmen, could be termed "firsts" in a long line of individually built precision reflecting telescopes. Science Museum, London. British Crown copyright.

6-in. aperture, one of which is shown in Fig. 3.10. Again it's of interest to note that this size, as that of Hadley's a half century earlier, coincides with the present popular amateur's 6-in. reflector. One of these instruments of 7-ft. focal length was matched against a larger reflector of about 9-in. aperture made by the famous James Short. It is said that Herschel's smaller instrument far excelled Short's. Thus, Sir William Herschel by long hours of painstaking effort and a delicate touch had become a master telescope maker. As a type of reward for his persistent efforts, Herschel in 1781 discovered a planet circling our sun out beyond Saturn's orbit, the planet now known as Uranus. The world was electrified and Herschel became instantly famous. Now with financial support he made a 12-in., an 18-in. and finally a 48-in. mirror having a 40-ft. focal length; truly a remarkable feat for the time, when we consider the cast speculum metal mirror blank alone weighed over a ton. Amateurs will still encounter Herschel's name since it is attached to a small clear glass plate, called a Herschel's wedge, that he developed to reduce the sun's light and heat for safer viewing. Later, William Parsons (Lord Rosse), after four casting failures, made the largest speculum metal mirror ever made, six feet in diameter and weighing four tons, called the "Birr Castle Giant" of Parsonstown, Ireland, as shown in the engraving of Fig. 3.11, an achievement for which he is justly well remembered. But plagued by poor location and mounting difficulties, Lord Rosse's primary astronomical contribution with this leviathan was limited to the revelation of the spiral structure of certain nebulae.

3.11 – Lord Rosse's six foot "Giant of Birr Castle." Its four-ton metal mirror was the largest ever cast. Note the figures for scale. From G. F. Chambers' Descriptive Astronomy (1868), **courtesy The Clarendon Press, Oxford.**

3.12 – William Lassell's 48-in. equatorial reflector on the island of Malta. Note the rotatable lattice tube and observer's cage 40 feet up on an elevator tower moved around on a circular railway. An assistant drove it by turning a crank timed to a clock's beat. From original engraving in the Memoirs of The Royal Astronomical Society, 1867.

Oddly, these earliest large instruments were often on crude altazimuth mounts that were difficult to maneuver. Possibly the mechanical or financial problems of superior equatorial designs for these giants were then insurmountable. Herschel was most productive with his smaller, more refined instruments and saw the polar caps on Mars, counted and studied the satellites and surfaces of Saturn and Jupiter, and noted the double stars. He was a tireless searcher for nebulae and increased the number of recognized nebulae from a hundred to several thousand. On top of all these accomplishments he made and sold many superb telescopes, reputedly high priced but undoubtedly well worth the amount asked.

During this period occurred "the battle of the giants," according to author G. E. Pendray; one where the goal now became that of larger and larger mirror diameters instead of the earlier goal of longer and longer tubes. While this era produced some giant "white elephants," big, awkward and useless, others made astronomical advancements otherwise impossible without enormous light-gathering power. Equatorial mounts came into use, some with many refinements. Astronomers now moved up from their cramped neck-wringing positions on the ground to become aerial artists, truly risking life and limb atop swaying cages on slings, winding stairs, tall towers and elevator platforms, to reach the eyepieces high up on the sides of these giant tubes (Fig. 3.12). Such rigors were faithfully endured, since this was the serious business of gathering more and more light, with each larger aperture often yielding new discoveries to thrill a now science-hungry world, one too long chained to the insignificant light-collecting power of the eye and hampered by the dim, hazy images at the bottom end of early refractors. Suddenly these large reflectors increased man's seeing power thousands of times with amazing results. Galaxies took form, planets revealed their surface structure and their swarms of moons, and lone stars suddenly resolved into sparkling close-coupled twins slowly swinging about a gravitational axis.

These great mirror telescopes had certainly come a long way from Newton's one-inch baby mirrors. While the reflector was from now on to stay ahead of the refractor in the battle for size alone, the refractor still remained doggedly in the race — the race of overall performance and service.

Quietly, around 1780, a Swiss bell maker, Pierre Louis Guinand, experimented with making optical glass and developed techniques that produced flawless glass in discs up to 6 inches in diameter, most particularly the much-needed high-quality dense flint glass. The refracting telescope was about to march again. Guinand joined Joseph von Fraunhofer, the inventor of the spectroscope, at Munich, where Fraunhofer now learned how to make fine optical glass under supervision of a master at the art. Fraunhofer then proceeded to design a new set of curves for the surfaces of an achromatic lens differing slightly from Dollond's. In this so-called "air-spaced" lens, the front surface was strongly curved and the central two required small, thin spacers at the edge to keep the lenses from striking at the center. The back was almost flat. (See Fig. 3.1b.) Although Fraunhofer died quite young, so superb were the corrections of lenses with these curves that any sized refractor could be made within the limits of glass discs available, and refractors to this day are made using Fraunhofer's design. These events brought the refractor firmly and permanently back as a large, practical, and convenient telescope, considered by many as better suited for planetary studies and certain specialized research than reflectors.

As telescope making swung to America, men such as Holcomb, Fitz

3.13 – **The mighty 40-in. Yerkes refractor points its great lens at the skies through the dome at Williamsbay, Wisconsin. This lens by Alvan Clark and Sons represents possibly the largest practical refractor lens size. Alvan Clark and Carl Lundin** (inset) **are posed beside the 40-in. lens elements and cell. Yerkes Observatory photographs.**

and Peate made fine optics, but their names sank into obscurity. Meanwhile at Cambridgeport, Massachusetts, Alvan Clark struggled for many years to polish refractor lens surfaces of a quality superior to those from the famous European makers. Clark was known to coat his fingertip with polishing rouge for a "touch up" of a local area of the surface to correct for a slight deviation in the glass's uniformity. Most likely the recognition of the unexcelled quality of this modest New Englander's efforts for perfection by an English astronomer, Reverend W. R. Dawes, widely known for the resolution limit formula under his name, started the firm of Alvan Clark and Sons safely off to success not enjoyed by earlier or contemporary workers. Reverend Dawes personally purchased four of Clark's lenses, and his enthusiastic comments on these left no doubt whatever as to their superb quality. Refracting telescopes continued to grow, possibly to maturity, with the great 40-in. discs of glass, polished to perfection by Alvan Clark and Sons, mounted on the long giant tube in the Yerkes Observatory at Williams Bay, Wisconsin, which is the largest refracting telescope in the world and probably as large as is practical for a refractor. Since this instrument marks the end of a long journey, with many stops and interesting side visits, the mighty Yerkes refractor is shown in Fig. 3.13, appearing more like a giant cannon trained on the sky than the high precision optical instrument it is.

Over a half century had to pass before the reflector grew beyond Lord Rosse's four-ton 72-in. metal leviathan. Actually a thin film of silver, less than a thousandth of an inch thick, revived the great reflectors. Two men, German and French physicists, Dr. Karl August von Steinheil and Jean Bernard Léon Foucault, independently learned how to lay down a highly reflecting thin film of silver on glass telescope specula. Most amateur telescope makers think of Foucault in connection with his famous "Foucault's Test," a method that permits the amateur mirror maker to rapidly shop test his telescope mirror to see if it is a perfect paraboloid as required for optical excellence. In America John Alfred Brashear not only made outstanding reflecting telescopes but furnished early amateurs a home silvering process for their glass mirrors.

3.14 – The 200-in. Hale reflector at Mt. Palomar represents man's greatest effort to date in the struggle for more light-gathering power. In 1948 the United States issued a blue stamp as a memorial to this feat. The observer now rides within the tube as shown, to control photography at the mirror's focus. Russia now has a 236-in. altazimuth reflector. Photographs from the Mount Wilson and Palomar Observatories.

3.15 — The world's largest solar telescope on Kitt Peak operates in an excellent location at almost 7000 feet. An 80-in. flat quartz mirror located 110 feet above the ground directs light down the shaft to a 300-ft.-focus 60-in. working parabolic mirror deep in the mountain to form an image almost a yard in diameter. Before judging size note the figure and car at ground level. Courtesy the Kitt Peak National Observatory.

Now with such beautiful reflecting surfaces and test methods available, only the size of glass discs that man could make, grind, polish, silver and mount would limit the reflector's progress. Soon 82-, 100-, 120-in. mirrors went into operation. Aluminum surfaces replaced silver. All these constant improvements culminated finally in the giant at Palomar, the 200-in. Hale telescope shown in Fig. 3.14, where the astronomer can now sit *inside* the tube right at the prime focus. A commemorative blue stamp was issued on this great observatory housing the world's largest reflector. In Fig. 3.15 we see a strikingly unique newer observatory totally devoted to solar research. A great quartz mirror at the top reflects the sun's rays down the slanting polar tube to place the sun's image in the laboratories below. This beautifully simple architecture deceives one as to size — note the auto and figure below.

All who build, assemble or buy an astronomical telescope are profiting from these many years of skilled work and intense rivalry of many men, which enlivened the history of telescope making. Telescopes have appeared on commemorative stamps from time to time (Fig. 3.16). Those who own even the most modest instrument worthy of the name of "telescope" have a possession that would have been truly priceless to famous early astronomers as Copernicus, Kepler and Tycho Brahe, who painstakingly studied the skies without optical assistance. Figure 3.17 spans the optical telescope age — from Tycho Brahe to the giant radio telescopes now searching the

3.16 – **Astronomical telescopes themselves have graced several commemorative stamps.** (left) **An azimuth refracting telescope appears on a dark blue-green 1949 issue marking the 50th anniversary of a Japanese Latitude Observatory.** (center) **Twin equatorially mounted refractors at solar work appear on this dark brown on yellow ochre 1957 Russian issue.** (right) **The 28-in. reflector at Sternberg appears on the rust brown and dark blue issue commemorating the 1958 International Astronomical Union Meeting at Moscow.**

3.17 – **Our review has spanned the optical telescope age from Tycho Brahe** (left) **and his great mechanical measuring instruments before lenses to the giant radio telescopes** (below) **now scanning the skies. The reddish-brown 1946 Danish stamp commemorates Brahe's magnificent works. In 1958 Haiti issued a striking bright red on deep blue stamp depicting the great Jodrell Bank English 250 foot steerable parabolic antenna, and in 1963 France issued the brown on sky-blue stamp with its great horizontal radio telescope receiving signals from a spiral galaxy.**

sky and recording distant catastrophic explosions that stagger the imagination.

For those interested in pursuing the history of telescopes in more detail I'd like to recommend a few of the books used as references. I was impressed by the style of the small book *The Telescope* by H. E. Neal (Julian Messner, Inc.), and the completeness of *The History of the Telescope* by H. C. King (Sky Publishing Corporation). Then there is the entertainingly written, justly praised work of G. Edward Pendray, *Men, Mirrors and Stars* and the classic *The Telescope* by Louis Bell (both out of print; see your library). Also two excellent, elegantly illustrated newer works, *The Picture History of Astronomy* by Patrick Moore (Grosset & Dunlap) and *Astronomy* by Fred Hoyle (Doubleday & Co.).

More detail on telescopes and observatories can be found in *Telescopes,* edited by Thornton and Lou Page (Macmillan) and *Explorer of the Universe* by Helen Wright (Dutton), a biography of George Ellery Hale and his works, culminating in the 200″ Hale telescope at Mt. Palomar. Below are some new ideas in observing and observatories.

3.18 – (left) **Most of us like comfort when observing. I derived much pleasure in "sky sweeping" with the counterbalanced observing chair shown. Hand cranks at fingertips easily control circular, vertical and horizontal motion. You are actually inside an altazimuth mount! Binoculars are turret 12, 20 and 40 x 80mm. Carl Zeiss and scope is a zoom eyepiece 4-in. refractor. Refurbished from Dr. E. Everhart's comet sweeper. See** Sky and Telescope, **Jan. 1974 for details of Mr. P.T. Menoher's motorized chair. Photo by the author.** (right) **I've used simple "lift off" contoured scope canopy covers many times over the years. Pictured is an elegant contoured one by E. Ken Owen. The scope is on a raised deck. Turn to color section for superb blue-lighted view of his entire observatory and scope. See pg. 84 for closeup view of his outstanding craftsmanship in design, detail and chrome finish. Photo courtesy E. Ken Owen.**

4

SMALL, MEDIUM and LARGE TELESCOPES

GENERAL

What kind, what size, what price? These are usually the first questions asked when someone wants to purchase a telescope for skygazing. They can be best answered by asking yourself a few questions in return. Is it for yourself, a friend, a child or teenager, an adult, or for a club or school? Will it be for astronomical use only or for looking around the countryside as well? Portable or permanently mounted? For a first introduction to the sky or a more thorough look? How much do you wish to spend?

You've probably already fixed your needs in terms of the above questions on reading them. Let's simplify things a bit by dividing *all* telescopes in three groups: small, medium and large, as defined at the end of Chapter Two. A discussion of each group should provide many answers to the remaining questions you may have. Size is not just the lens or mirror size, as discussed, but is based partially also on type, bulk, ease of operation and portability. Please also note that true astronomical telescopes present an *inverted image* and require an accessory "erector" for any planned terrestrial use.

Commercial telescopes in each size group will be presented along with some suggestions and considerations on acquiring an instrument to meet your specific needs. Since it's impossible to provide fine detail on each instrument illustrated, pick one or more that seem to fit your needs and write the supplier for complete information. Names are provided in the "Supplier" directory at the end of the book, with manufacturers listed alphabetically according to key parts of their name. Why not contact several? Many manufacturers supply a broad line, and their literature can be both educational and helpful; however, temper your decisions with knowledge gained from material presented here or by tests described in Chapter Seven.

Naturally only a few instruments can be shown of all those available. It is believed the manufacturers listed provide quite satisfactory telescopes and are often quite justly proud of the instruments they build. Most makers constantly strive for improvement, so expect changes. I've known those presented for several years. One should purchase the very best (usually the costliest) he can possibly afford of any given type or size. Here, possibly more than with most merchandise, price is often a fair measure of quality. The old adage that "the best is none too good" certainly applies to optical quality. Beware of "bargains," since there is no form of optics that someone can't make a little cheaper — and a little worse.

Above all, ignore the "high power craze" in advertisements. In *all* telescopes the *usefulness* of powers much above about 50 times magnification for each inch of lens or mirror diameter (*i.e.,* 100X for a 2-in., 150X for a 3-in. and 300X for a 6-in., etc.) is open to serious question. While power claims above this are correct by definition, they are highly misleading in that such powers have little value and are rarely of practical use in observing, the objects seen being too dim and ill-defined. Such magnifications *per inch of aperture* are even less often used on larger instruments because of temperature and atmospheric disturbances.

SMALL TELESCOPES

Those of the astronomical type are essentially for youngsters, or for a "first look." This group comprises refracting lens telescopes up to 2½-inches in size and reflecting mirror telescopes up to 4½-inches in diameter. Lens diameter of small refractors is often expressed in millimeters — 25 millimeters (mm.) to an inch. Should it be a refractor or reflector? In "first" instruments the small refractor has many advantages, and few of the disadvantages that plague its larger brothers. It is highly portable, stays in adjustment, is easy to clean, and easy to point at objects without an accessory "finder" telescope. It is ideal for terrestrial viewing, the lower-powered scopes often having erecting lenses or prisms and the higher-powered truer astronomical types easily so provided for dual use. However, it is true that for a similar investment and *astronomical use only* by a mechanically inclined boy or girl, a 3-in. or 4¼-in. reflecting scope will permit seeing much more detailed and brighter astronomical images. The 4¼-in. is markedly superior to the 3-in., if you can afford it. Nevertheless, reflectors are considerably "fussier" to adjust and care for, and are not very satisfactory for terrestrial use. In summary, the small refracting or reflecting telescopes are most useful as introductory instruments for younger members of the family. High-quality "spotting scopes" as used terrestrially for target work, bird-watching, etc., but which can still be turned to the skies, are completely covered in *Binoculars and All-Purpose Telescopes,* published by Amphoto.

Cost? Prices can range from $3.00 up to about $200.00. Prices presented *are approximate*—please check. Besides the book *All About Telescopes,* Edmund sells a small "assemble it yourself" kit to make an 8X scope (Fig. 4.1). You may wish to assemble one or more for the children. It's for the skies only — inverted image. The $20.00 Bausch & Lomb 10-power shown is complete and ready for all-around use either terrestrially or astronomically. It can be hand held or mounted on camera tripod with an adapter. The Bausch & Lomb 20-power shown is another dual-purpose telescope, and is quite useful for rifle target spotting. It should be used with a camera tripod, as should any instrument over 10 power. Bausch & Lomb also sells the elegant Balscope-60 with fingertip zoom focus and power control at a substantially higher price. Bushnell has introduced the small, economical, tripod-mounted 50 mm. zoom telescope with a 9 to 30 power range, a good first all-purpose scope for young children. Sky observers should get the 45° *inclined eyepiece* Balscope-60.

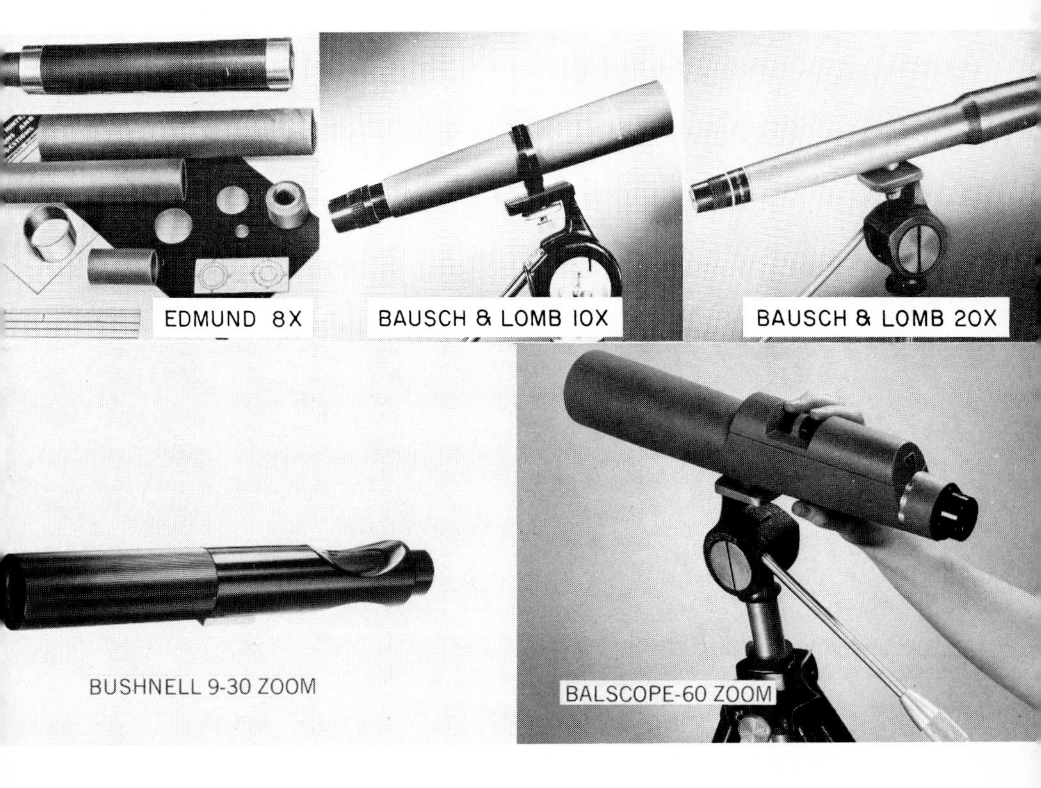

4.1 – Suitable small refracting telescopes for a youngster's first look at the skies, reasonably priced. All but the first can serve terrestrially too. Photos courtesy suppliers.

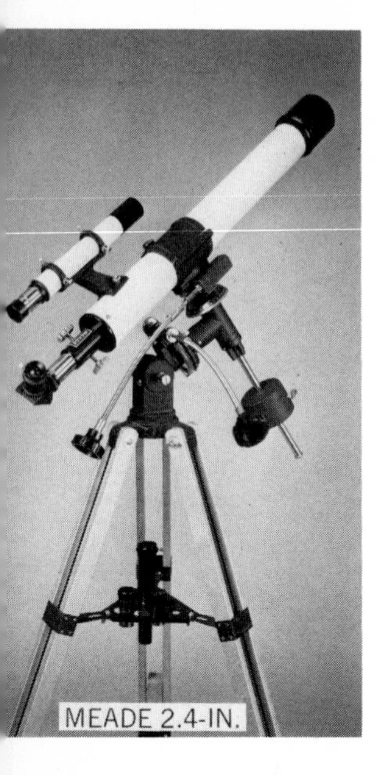

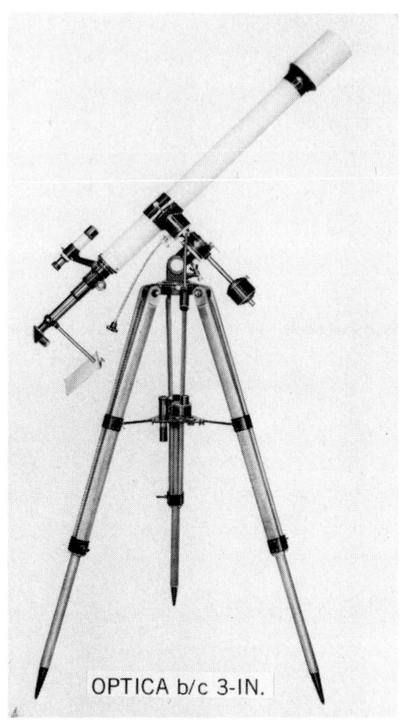

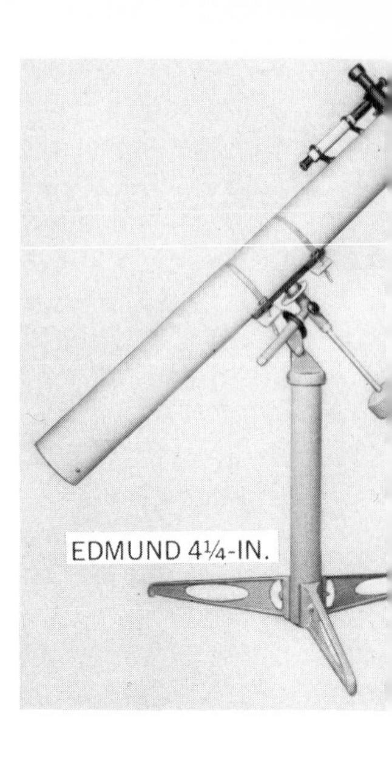

4.2 — Representative more powerful astronomical—terrestrial refractors and a small astronomical type reflector for the ardent beginner. Prices vary depending on accessory drives, eyepieces, erectors, etc. Photos courtesy suppliers.

A Bushnell Spacemaster, equipped with zoom and wide-angle eyepieces was selected for the joint USA-USSR Apollo-Soyuz successful test project — our first joint effort with the Russians, July '75. Scope unit length of only 11⅝-in., was an important factor in the restricted spacecraft quarters.

Let's investigate some more truly astronomical telescopes. The 2.4-in. Meade on the equatorial mount shown in Fig. 4.2 (left) is an excellent, top-quality, all-purpose small refractor for a serious young amateur. A slightly larger model (4.2 center) is well worth extra money. An economical triple-turret erecting eyepiece is available for these. In the line of reflecting telescopes I'd suggest never buying one having a mirror less than 3 inches in diameter, and I'd strongly urge going to at least a 4-inch one if you can possibly afford it. An economically priced 4-in. reflector by Edmund will permit a lot of "seeing." An excellent 4¼-in. model with a fine, rugged, conventional equatorial mount and drive is offered by Criterion. I consider it an excellent buy.

After you receive an instrument you may wish to consult Chapter Seven on testing. Also consult the directions that come with it, a procedure often successful when everything else fails!

MEDIUM-SIZED TELESCOPES

The medium-sized instrument is ideal for all adult or junior amateur astronomers. These telescopes range in size from 3- to 5-in. refractors to 6- to 10-in. reflectors. Most quality types are capable of excellent performance in detailed study of astronomical objects and are eminently suitable both for pure personal pleasure or for a student or serious amateur observer. The smaller sizes in this class are still reasonably portable, but the larger ones deserve fixed mounts. Many are suitable for small observing groups and serve quite well as portable teaching instruments. I believe with this size range, and larger, enough money is invested that the reader should write the manufacturer for the very latest prices and check for current model literature. It pays to shop a bit, but do buy the best you can afford. *All* appropriate equipment may be used for astrophotography with pleasing results (see Chapter Ten).

It is with this group of medium-sized instruments that the question of whether to buy a refractor lens telescope or a reflector mirror telescope is most often raised. In the smaller instruments the refractor is justly preferred, and in the larger group reflectors clearly take the lead. Because of the complexity of the refractor versus reflector story, let's postpone it until later to take a good look at *representative* medium-sized telescopes of both types generally available and become better acquainted with their general characteristics. We can then consider which type to buy, refractor or reflector.

The 3-in. refractor and the 6-in. reflector are by far the most popular medium-sized instruments. In Fig, 4.3 we see an economical 3-in. equatorially mounted Edmund. A 4-in. is also available. The 3-in. Unitron shown is a beautiful, top-quality precision instrument for which many useful accessories are available (see Figs. 5.4 and 5.6). A 4-in. Unitron appears in Fig. 3. Criterion makes the very practical, lightweight 6-in. Dynascope shown, as well as more elaborate models. Such reflectors are excellent examples of 6-in. models available from most makers in the $300.00 range, a size and price range in which best all-around buys for the money exist for the more serious young amateur astronomer or starting adult. Jaegers and Edmund offer "do-it-yourself kits."

In Fig. 4.4 we see an unusually portable 4-in. $f/15$ refractor, hence its listing with medium-sized instruments. You can make a fine, trouble-free,

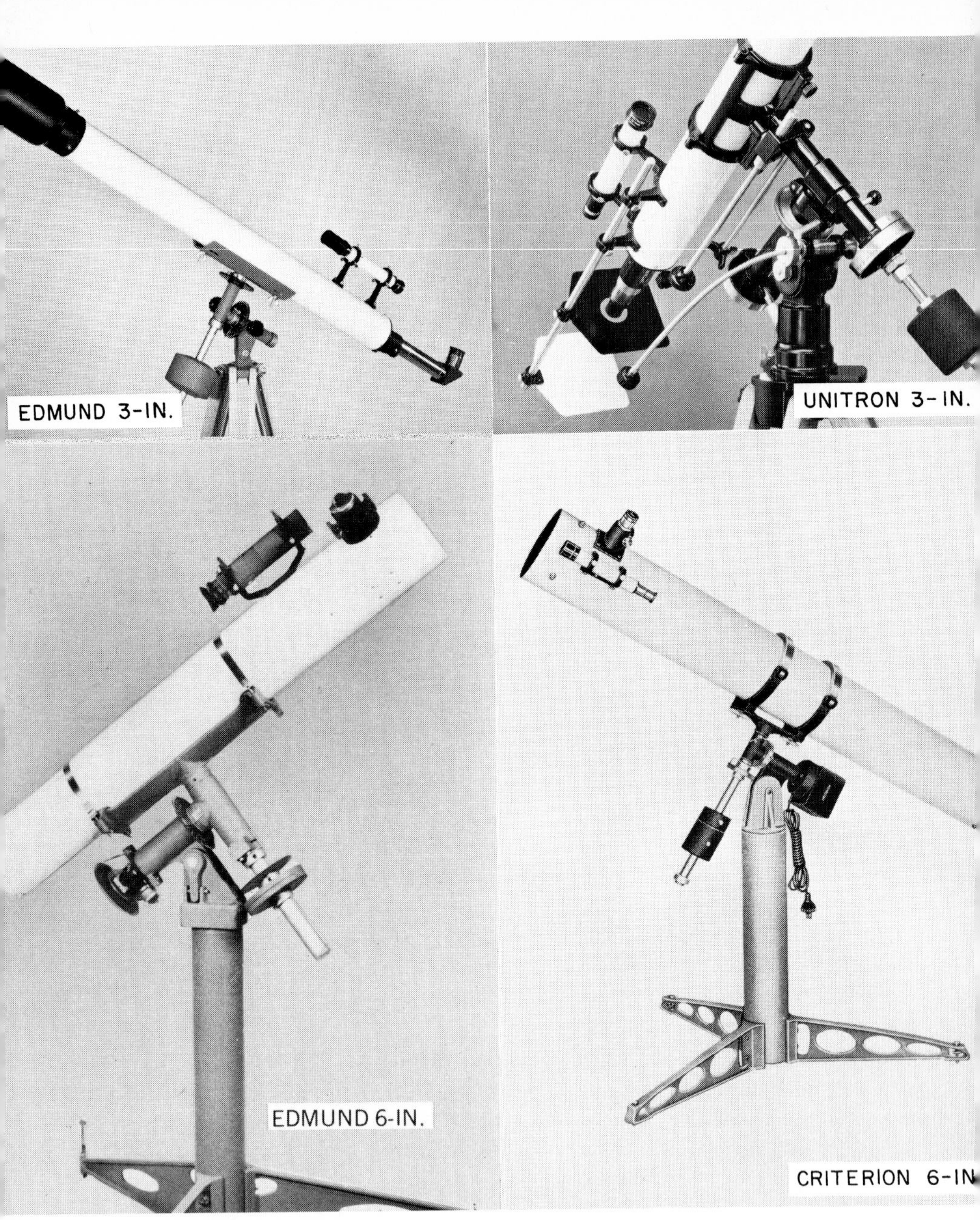

4.3 — **In refractor lens telescopes, the 3-in. represents the best all-around investment, while in reflecting mirror telescopes the 6-in. enjoys a similar role. Above are two representative examples of each. Ranging from $250 to $500. Prices vary with accessories. Photos courtesy suppliers.**

4.4 — **The short-focus 6-in. refractor and standard 8-in. reflector are growing rapidly in popularity. They can still be portable and are fine for group observing or serious study. Prices range from about $400 up. Check Jaegers' 6-in. f/10. Photos courtesy suppliers.**

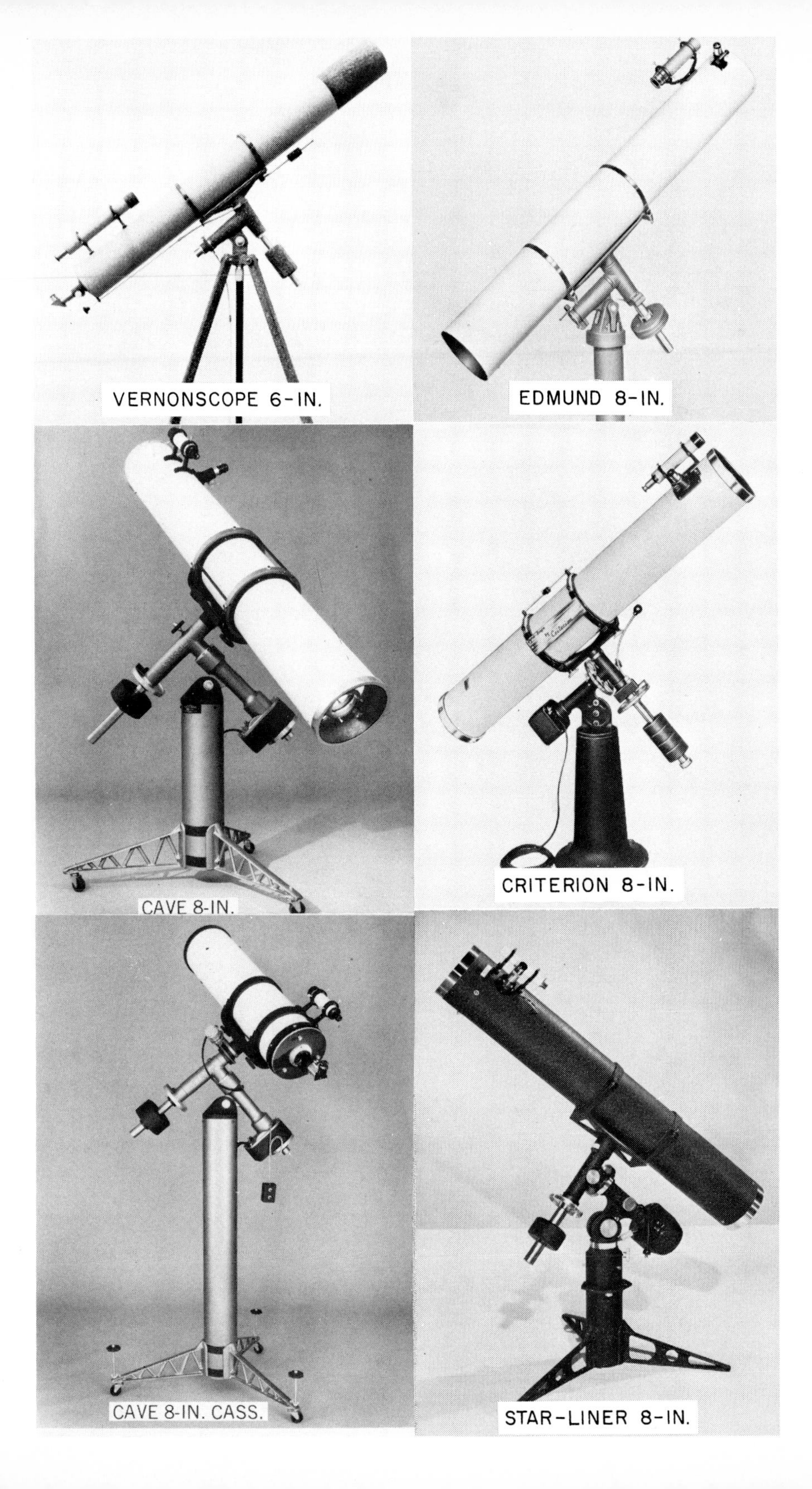

VERNONSCOPE 6-IN.

EDMUND 8-IN.

CAVE 8-IN.

CRITERION 8-IN.

CAVE 8-IN. CASS.

STAR-LINER 8-IN.

useful instrument, starting with a Jaegers 4-in. *f*/15 or 6-in. *f*/10 lens. Next in popularity to the 6-in. reflector is the 8-inch. Accordingly presented are fine models by Edmund, Cave, Criterion, and Star-Liner. Note also the fine 8-in. Cave Cassegrainian at the book front — Acknowledgment, Fig. 3. Each is well worth considering, and since you will be investing $300 and up why not obtain and carefully study manufacturer's literature? An 8-in. should have a fixed mounting and be housed. Most can be readily assembled and disassembled into three or four easily moveable parts, a method used by many to make them portable. In poorer climates or where rapid evening temperature drops occur, an 8- or 10-in. reflector does not have the advantage over the 6-in. one might expect, and the *very best obtainable* 6-in., with a full complement of accessories, including orthoscopic eyepieces, star and solar diagonals, etc. might serve better for a similar investment.

In Fig. 4.5 (left) is shown a new Cave super deluxe 10-in. portable Model "C" with circles, drive, fully rotatable tube, tube counterweight and an elegant finish of chrome plate plus polished aluminum. It's available on pier for observatory use. Cave has long been known for superb optics. Cave also makes fine 10-, 12- and 16-in. Cassegrainians similar to the 8-in. model shown in the front of the book. Actually a 12-in. Cassegrainian, because of its compactness, could be considered a medium-sized instrument; accordingly, at the right of Fig. 4.5 is a cleanly designed one by Star-Liner, with full accessory equipment. Shown on an observatory pier, it's also made as a transportable model as is Celestron and Criterion's Dynascope.

REFRACTORS VERSUS REFLECTORS

One of the most heatedly controversial questions is whether you should acquire a refractor or reflector telescope. While it's a relatively bloodless battle, this war between lens and mirror telescopes has raged since Newton's first reflectors, as related in Chapter Three. Most of the arguments have about the same character as those revolving around which is preferable — American cars or imports, automatic or stick shift, or married or single life. Actually there is no absolute answer — too much depends on an individual's interests.

Why not study some advantages and disadvantages of each telescope type and then choose one to suit your particular needs. To do this we must first consider your own special interests.

Primarily interested in astronomical use only? — or want terrestrial use as well? Fixed or portable instrument, and if portable how portable?

Are you "mechanically inclined?" Do you like to fix and adjust things or would you rather observe only?

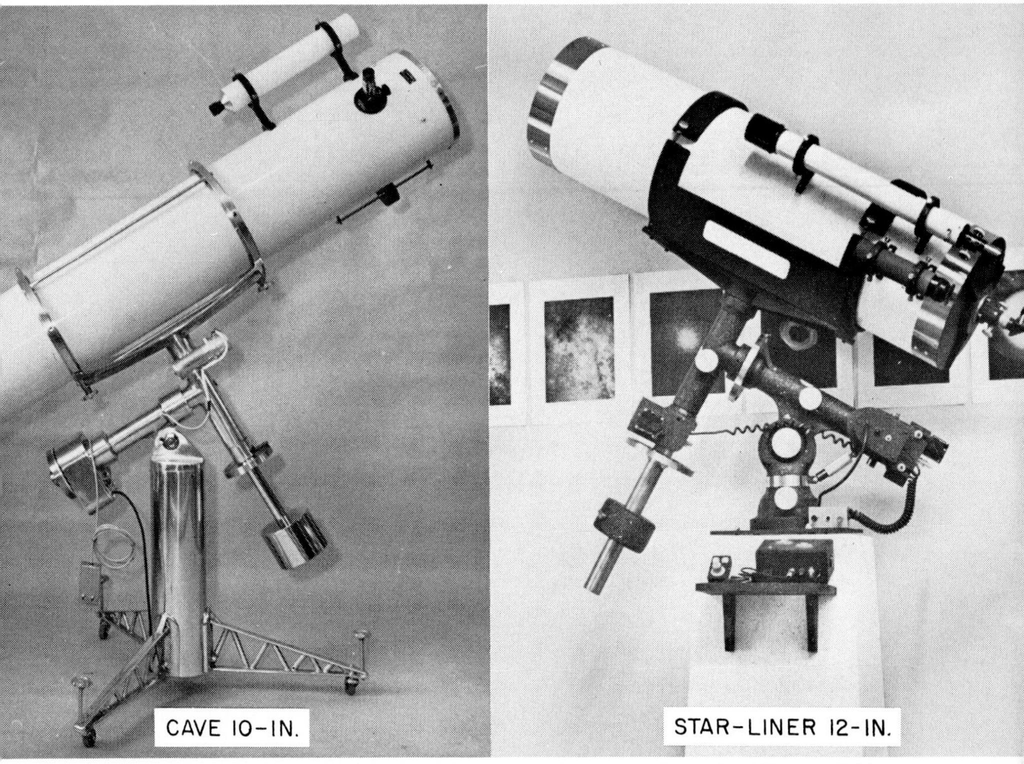

4.5 – **The 10-in. Newtonian** (left) **or the 12-in. Cassegrainian** (right) **can be either portable or observatory mounted. The Newtonian form is usually less costly, the Cassegrainian more compact. Each is capable of excellent performance. Photos courtesy of suppliers.**

Are you interested in general all-around observation or do you have special areas of interest, as the planets, moon and sun; or stars, double stars, star clusters and dim nebulae?

Do you plan astrophotography?

Do you have a good observing climate or possibly one with rapidly changing evening temperatures, heavy dewing, etc.?

With these questions in mind let's present the "pluses" and "minuses" of refractors and reflectors in the briefest possible form, starting with reflectors first because of their great popularity.

REFLECTING TELESCOPES

Advantages

Achromatic — color free — that is, no extraneous color in image.

Low cost — per inch of aperture. A big "plus" for reflectors.

Compactness — in 6-in. and larger sizes.

Convenient — Eyepiece location good; mounting compact for medium sizes.

Excellent for star and nebulae work where resolving power or great light grasp is essential.

Excellent for astrophotography.

Disadvantages

Temperature effects — open tube and changing temperature comparatively adverse for reflectors.

Adjustment and its maintenance requires reasonable interest in mechanics. *Delicate* mirror surface.

Mirror accuracy — single mirror surface *must* be of highest perfection for optimum performance.

Obstruction and scattering of light by the diagonal and spider reduces performance — particularly planetary detail.

REFRACTING TELESCOPES

Advantages

Ideal modest sized telescope, particularly in 3- and 4-in. sizes.

Terrestrial use — easily converted by erecting eyepiece units.

Maintains adjustment — optics tend to stay in collimation, etc.

Upkeep is essentially zero.

Closed tube — minimal temperature change effects.

No central obstruction — permitting unusually good performance on plentary, lunar and solar observation.

Disadvantages

Color error — slight secondary image color, increasing rapidly with size.

Cost high — per inch of aperture.

Long tube — per aperture — transporting and housing problems in larger sizes.

Mounting — a "gawky" tall mount. Eyepiece location poor without star diagonal.

Photographic use requires filters and relatively bright objects.

Disadvantages all increase rapidly with size — i.e., color error, portability, housing, etc.

Comments: Broadly speaking, a reflector costs about half as much as a refractor for the same aperture and quality, hence we readily see that the reflector gives the most astronomical "observing potential" for the money. It presents objects in their own true natural colors. Medium-sized refectors (6-in. $f/8$ to 10-in. $f/6$) are compact, easily mounted and moved, and have eyepieces in good observing positions. But reflectors are a bit "fussy" to adjust and to keep adjusted for the mechanically non-inclined. Performance can be adversely affected by rapid temperature changes in some locations. Reflectors are good for *all* observing but best with stars and the dim sky objects, where the large aperture (for the money) and the fast *f* value (*f*/8 against *f*/15 for refractor) provide resolving and light-gathering power desired. They are fine for narrow field astrophotography.

Hence we can say in summary the medium-sized refractor is a fine, all-purpose, trouble-free telescope requiring little care or adjustment and performing comparatively better than a reflector under adverse temperature changes. However, it costs at least twice as much as a reflector of equal aperture and quality. The advantages resting with the refractor diminish rapidly as we move from 3- or 4-in. up through 6- and 8-in. sizes. The absence of central obstruction permits the refractor to perform unusually well on planetary observation. However, don't forget that a 6-in. reflector, at half the cost, will under good conditions perform as well as a 5-in. refractor and have a marked advantage for photography.

After considering the questions at the start of this section and the comparative information, you have the wherewithal to make *your* choice.

LARGER INSTRUMENTS

Six-inch or larger refractors and 12-in. or larger reflectors are powerful observational–photographic instruments, ideal for clubs and schools. They deserve the best of attention and housing. A select few from recognized companies are illustrated to show the fine instruments now available. The values received in true utility for the money now invested is quite phenomenal when we consider what similar performing instruments, particularly from the continent, cost not too many years ago. Modern accessories in the area of drives, guiding controls, solar filters and photographic accessory equipment add even more to their precision, usefulness and convenience.

In these larger instruments the reflector is "king" for both value and utility. The "folded" optical systems, as the Cassegrainian and many related modifications, permit a comparatively large aperture instrument of 20- to 36-in. to be housed in reasonably economical observatories of utility types. No longer need devoted groups, smaller schools or colleges be without the finest facilities if an *enthusiastic group* (or even a single devoted individual with means) sets its collective mind on a specific accomplishment. Refer to Cave, Celestron and Criterion.

Six-inch short-focus (*f*/10) or standard long-focus (*f*/15) refractors can be assembled from components from A. Jaegers and The American Science Center, etc., the short-focus one making a fine, reasonably portable teaching instrument — with some slight reservation by critical planetary observers who prefer *f*/15 refractors. A standard 6-in. refractor takes quite an observatory, yet has some merit as a remarkably trouble-free instrument for continuous long service — a factor of importance in certain educational activities. Larger refractors should have individual attention of qualified persons wherever they are used and are not considered in the present work. Their increasing color error with size should be recognized. It does seem, though, that an observing group of limited means should consider obtaining one of the quality 8- to 12-in. refractor lenses by established early makers that often gather dust on shelves of institutions or museums, and construct a "folded" instrument along the lines discussed in Chapter Six.

Unquestionably the standard 12-in. Newtonian of *f*/5 to *f*/7 ratio is the all-around "best buy" today in larger equipment. The excellent 12½-in. reflectors by several makers are almost identical in design and appearance to their neat, smaller 8-in. counterparts already illustrated. The Criterion deluxe Dynascope has convenient dual finders. A standard 12½-in. Cave is nicely illustrated in a home observatory in Fig. 2.5.

Figure 4.6 presents four larger instruments; each one is representative of a fine line of telescopes. The new 5- to 14-in. Schmidt-Cassegrainians are excellent.

While the Cassegrainian form provides a shorter tube length, I'd tend to consider its merits for observatory use only in sizes *above* 12-in. aperture. Even though I own a fine 12-in. Cassegrainian, I lean towards a 12-in. *f*/6 or *f*/7 Newtonian for all-around use. It has been stated by noted astronomers from Newton's day on down that Cassegrainians have all of the faults of most reflecting telescope systems with the merits of none — a statement most likely sparked by the instrument's large, light-scattering secondary mirror. This possibly overharsh statement has an element of truth, but it does not recognize compactness, beneficial eyepiece location and *portability* of 8- to 12-in. sizes, or smaller observatory requirement for larger ones; and the wonderful freedom of color error in so short an instrument.

SPECIAL FORMS

I'd like to present a few of the interesting instruments that differ from standard forms. There is a telescope designed for viewing ideally the wide fields of star clouds and dim sky objects, the Richfield Telescope, or RFT. It shows the greatest number of stars possible with its aperture and presents splendid wide-angle views of the Milky Way and galactic constellations not possible with conventional instruments. The 4-in. or 5-in. RFT is only about 20 inches long and cradles nicely in the arms for hand holding. A homebuilt one is shown in Fig. 6.2. Tuthill makes a fine small 4-in. RFT. See Fig. 4.7 (upper right). Cave, Jaegers and others make RFT's. A fine 5-in. *f*/5 RFT refractor with angle eyepiece in a yoke mount is ideally suited for wide-angle low-power work and most convenient to use.

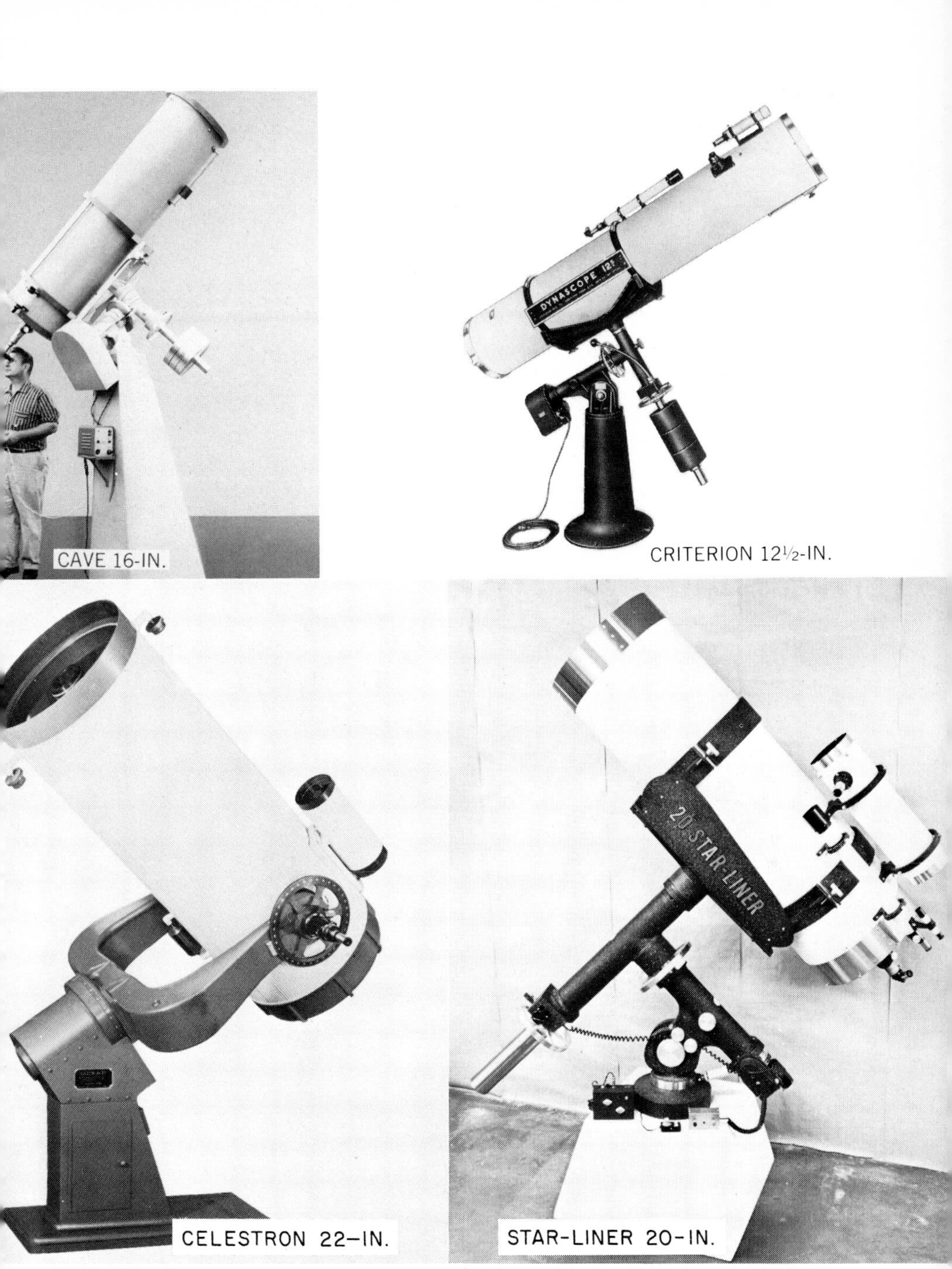

4.6 – **The reflector is the preferred larger-aperture instrument. Shown are four excellent examples of larger ones ranging from 12 to 22 inches, each quite suitable for large groups or schools. Excellent optics are now available in any size. Photos courtesy suppliers.**

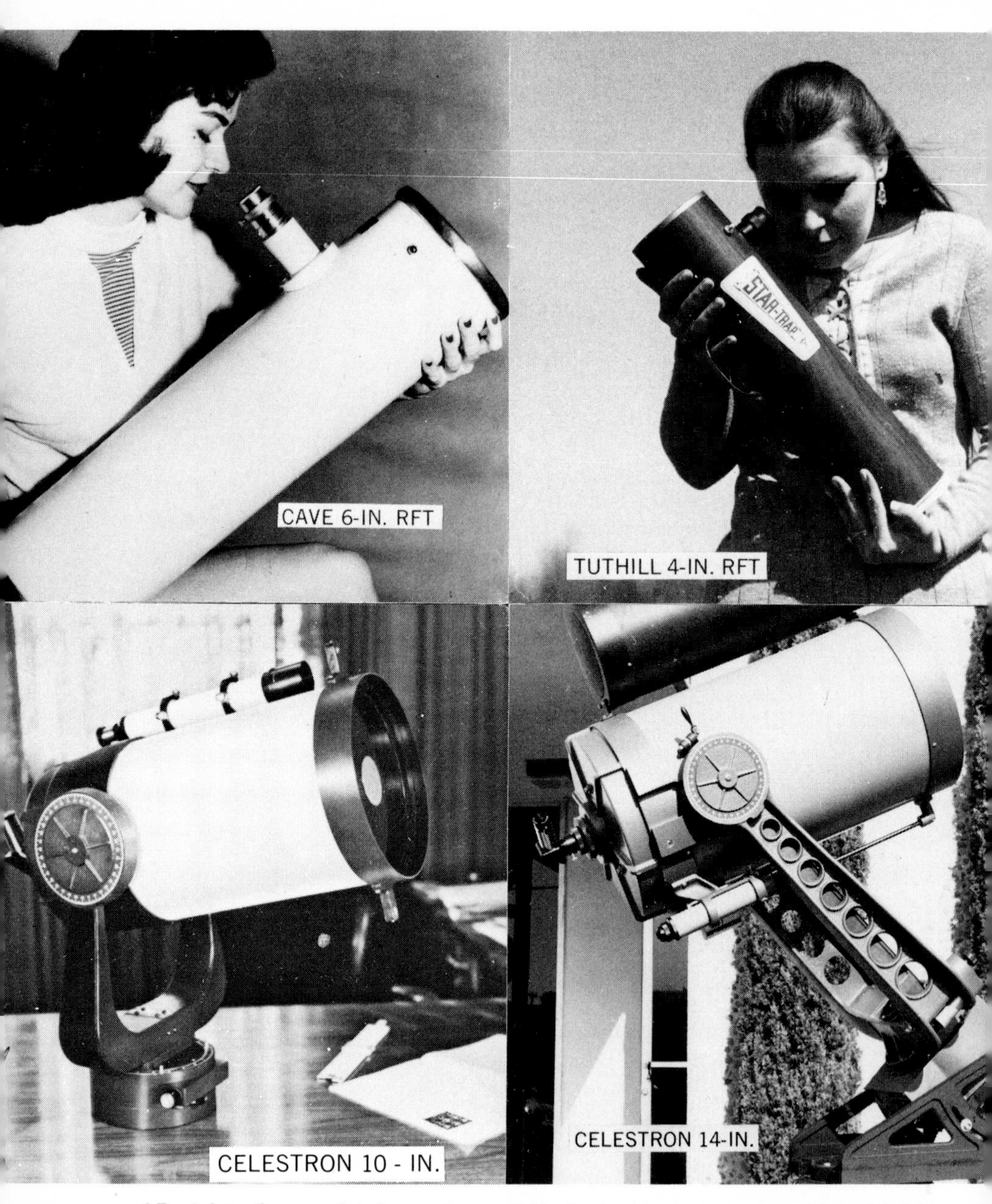

4.7 – **Interesting special forms.** (upper left) **A short-focus low-power 6-in. Richfield Telescope (RFT). It may also be hand held.** (upper right) **A new "closed tube" 4-in. RFT—a best buy.** (lower left) **An ultracompact 10-in. Schmidt–Cassegrainian form.** (lower right) **A 14-in. Celestron "transportable" Schmidt–Cassegrainian. About as large as one really cares to move.**

In lens-mirror catadioptric systems the superb 3½-in. lens-mirror Questar has been previously mentioned and is well shown in Fig. 2.6. In Fig. 4.7 (lower right) we see a 14-in. transportable Schmidt-Cassegrainian made by Celestron — extremely versatile yet portable. It focuses from only 500 feet to infinity by a convenient dial that shifts the small secondary mirror. In the larger lens-mirror systems we find many candidates. Celestron-Pacific makes the Schmidt-Cassegrainian form in all standard sizes from 5 to 36 inches in aperture. The 5- and 8-in. models are the most practical for the average amateur. This optical design is capable of exceptionally good off-axis definition compared to the more common types. Shown (lower left) is Celestron's earlier exceptionally compact 10-in.

Cave Optical Co. manufactures an unusually large (6-in.) RFT that can be readily hand held. See Fig. 4.7 (upper left). RFT's should have tripod mounts, since I recommend a steady support for sharp area viewing.

4.8 – **This 4-in. equatorially mounted Unitron refractor carries a rather full quota of accessories, including a drive unit, finder and guide scopes, turret eyepiece holder, slow motions, and cameras for both wide-angle and high-power astrophotography.**

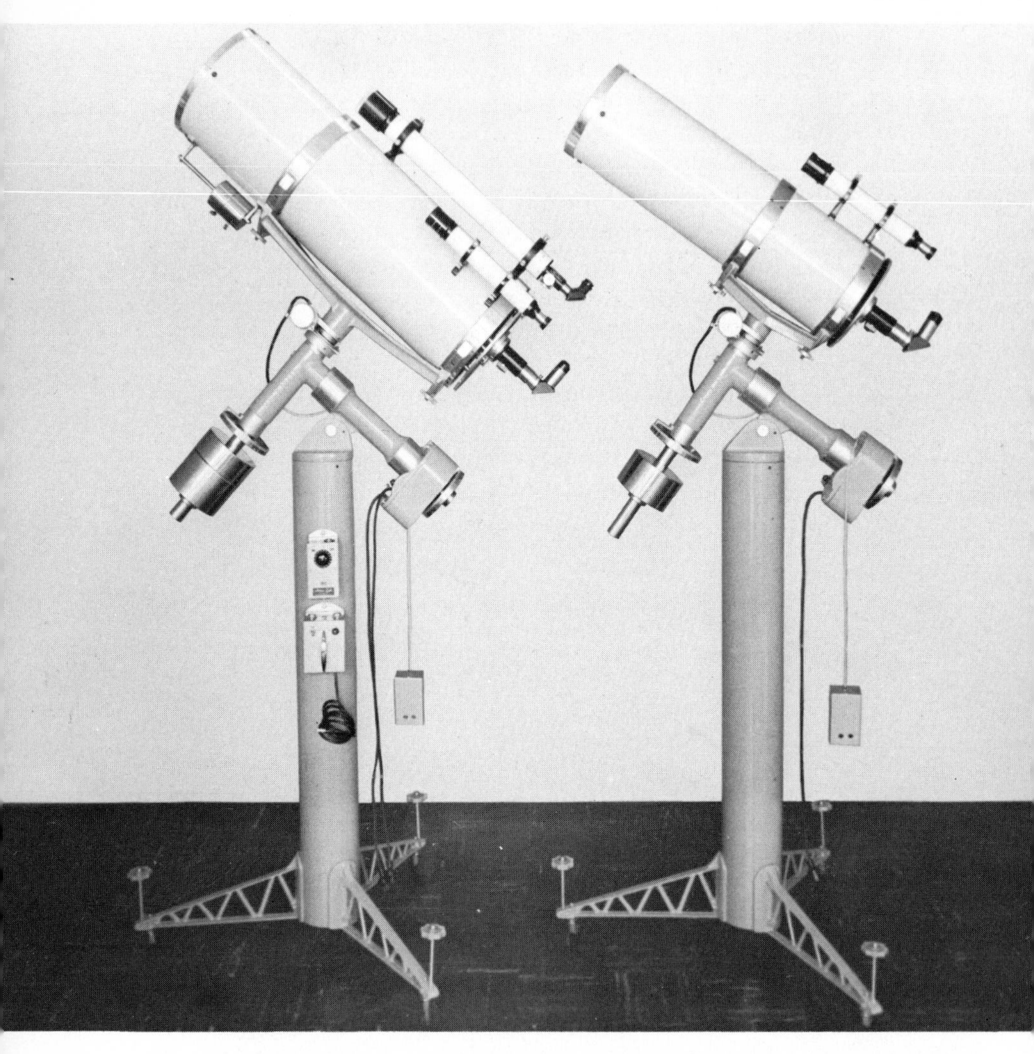

4.9 — A pair of cleanly designed, well-made reflecting telescopes by The Cave Optical Co. Each is fitted with a convenient slip-ring sidereal drive. Such compact 8-in. and 10-in. Cassegrainian designs maintain a high degree of portability for such relatively large apertures. More economical Newtonians may be had on these mountings.

5

USEFUL ACCESSORIES

EYEPIECES

The use of wrong eyepieces has prevented the good performance of many telescopes, especially reflectors. In fact, I've seen superb commercial optics used with cheap eyepieces of ancient design that reduce their potential markedly, a ridiculous state when we think of the minor cost to improve the system. Fig. 5.1 presents designs of some of the most used eyepieces and Fig. 5.2 illustrates representative examples. The *Ramsden* form shown at (a) is probably the most economical of the types worth considering. It is a so-called *positive* eyepiece, since the image plane (at arrow) is in front of the eyepiece unit, making it a good form where cross hairs or reticles are required at the diaphragm stop indicated. The characteristics of the Ramsden make it good for reflectors, but preferably not for aperture ratios much faster than the common $f/6$ to $f/8$ ratios. The inexpensive *Huygens* (d) had best be confined to normal $f/15$ refractors or similar small-lensed instruments. Similar microscope eyepieces have been used by beginning amateurs, who often were quite happy until they looked through better ones. The Huygens is a *negative* eyepiece, the image plane being *between* the field lens and the eye lens, as shown by the diaphragm stop and arrow in the figure. The *Kellner* (b) has an achromatic eye lens unit and may be considered an improvement over the Ramsden. It is commonly used in standard field width binoculars, and is available to many for astronomical use at economical prices through surplus sales. The eyepiece at (f) consists of a pair of achromats arranged as shown and often of identical construction, yet each member may be quite different to yield even orthoscopic characteristics. Vernonscope revived the superb Brandons (50° orthoscopic field, parafocal, rubber eyecups and screw-in filters) — in my opinion the best of all eyepieces. The true

Zeiss design *orthoscopic* (e) consists of a cemented triplet field lens appropriately coupled with a single eye lens. This excellent eyepiece type has a moderate field of about 40°, but covers beautifully from edge to edge. It is an excellent choice for fast aperture ratios, as $f/4$ systems, etc. There are numerous variations in orthoscopic designs. The *Erfle* design shown at (c) is a true "wide-angle" eyepiece, essentially as introduced by Dr. Erfle some years ago for binoculars. It should contain no fewer than three lens units, at least two of which should be cemented doublets, though all three may be. Some "super" wide-angle eyepieces may have four or more elements. It should be used in "Richfields" or wherever the widest possible view fields are desired. The large surplus 1¼-in.-focus, three-achromatic-lens, aluminum shell eyepiece shown in Fig. 5.2 is a remarkably good buy. For its full 70° field, a 1⅝″ diameter eyepiece tube must be used.

The "zoom" eyepiece shown is now widely available, having constantly variable power based on a constantly variable focal length, usually about 8 to 21 mm., and the focal length of the mirror or objective it's coupled with. As it is now made, I've found it fine for planetary and narrow field work, with possibly a touch of added color. Edge of field definition suffers at some settings and eyeglass wearers may resent the restricted (to them) view field. The casual observer might prefer this and a 2X Barlow over a wide range of eyepieces, but it is not likely to please the fastidious observer. Fields given are *apparent*. For *real* view field, divide by instrument's power.

Not illustrated but subject to consideration, particularly for planetary observation, is a solid single-unit eyepiece of cemented construction. The Zeiss monocentric cemented three-element eyepiece (East Germany) is possibly the best, covering about a 30° angle. The widely listed Hasting's "triplet" magnifier makes a fine narrow field eyepiece (± 20°) as does a simple small achromat (±10°) used with the most curved surface facing the eye. Such solid oculars are remarkably free from "ghost" images and light scatter, hence are good for planetary work, where their narrow view angle is of little concern. Clavé (Paris) makes fine Plossl types.

All eyepieces should be "optically coated" with so-called non-reflecting surfaces. This is particularly important in certain designs, as Kellner types, particularly early ones. Some American-made orthoscopics have triplet elements with equal outer curves, instead of the differing curves of the original (Zeiss) form — probably to reduce manufacturing costs. These have bad "ghost" reflection images unless coated. Since the coating on the outer surface facing the eye is easily damaged and can look "moth eaten" and esthetically bad, some firms prefer not to coat this surface. Either way makes little difference in actual performance. Good hard coats resist reasonable and *gentle* cleaning.

The Barlow lens, often termed an amplifying or magnifying lens, should

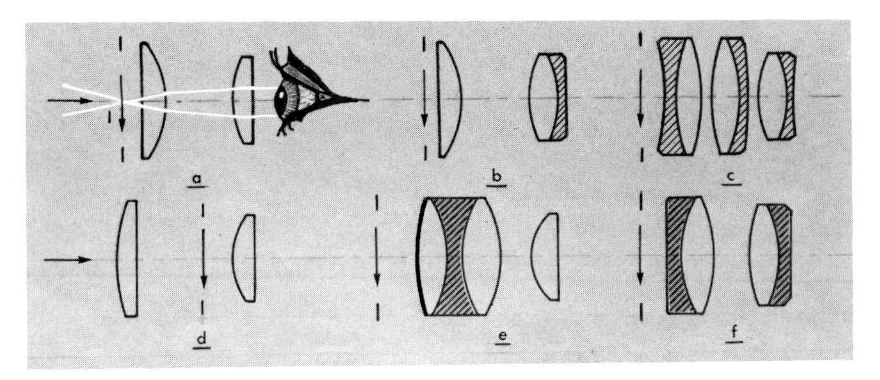

5.1 – Eyepiece designs of the six most commonly used forms. As presented, the Ramsden, Kellner, Erfle, Huygens, orthoscopic (Zeiss type) and Plossl (gunsight) eyepieces. Drawing by the author.

be more widely used. Its greatest advantage is to eyeglass wearers when higher magnifications are wanted. The added color error of good achromatic Barlows is practically nill. The 2.4X Vernonscope Dakin Barlow is the best, but *not* on single negative lenses. A Barlow of 2X or 3X amplification allows the use of medium-power eyepieces in which wearers of glasses can see the full field of view. If one selects the right medium-power eyepieces, as 12, 16 and 20 mm., a 2X Barlow yields the equivalent of having 6, 8 and 10 mm. high powers *with* good eye relief. In Fig. 5.3 we see how the Barlow works effectively to increase the apparent focal length and aperture ratio. In the figure, (p) is the original focus and (p^1) the new focus. The magnification (M) equals distance (C) divided by (A). For a given focal length Barlow lens (FL) anything you wish to know can be calculated from the formulas

$$M = \frac{FL(M\text{-}1)}{A};\ \text{or}\ A = \frac{(M\text{-}1)}{M} \times FL;\ \text{or}\ C = (M\text{-}1) \times FL.$$

5.2 – An ideal eyepiece set. Focal lengths of 8, 12, 20 and 32 mm. (wide-angle with adapter), 2X Barlow and 8-21 mm. zoom. Used with a standard 6-in., 48-in.-focus reflector these provide powers of 38, 60, 75, 100, 120, 150, 200 and 300X, with continuous zoom from 60 to 150X, or 120 to 300X. Photo by the author.

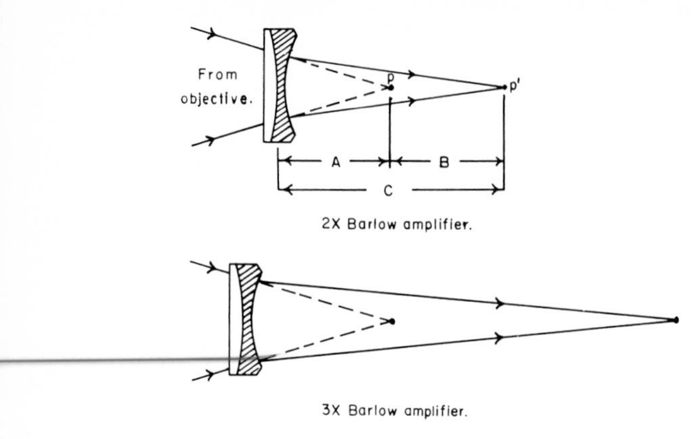

5.3 – How the Barlow amplifier lens works. Relationship of initial and final focal points for 2X or 3X amplification is illustrated. See text for means of calculating powers, focal distances, etc. Drawing by the author.

The eyepiece, or camera, is always moved farther back by the distance (B), which becomes quite great for high amplification.

DIAGONALS, ERECTORS & FILTERS

A *star diagonal* makes the difference between comfortable and uncomfortable observing with a refractor. As shown in (a) and (b) of Fig. 5.4, a mirror or a prism unit inserted into the telescope's eyepiece holder diverts the light near focus at right angles up to the eyepiece in the unit's holder. One can rotate this unit about to view at any desired comfortable angle, instead of at neck-kinking angles. When you look downward, the image is erected top to bottom, but remains reversed right to left. Either 90° prism or first-surface mirror diagonals of good quality are very satisfactory for ordinary refractors, the Zenith prism type requiring less care. The mirror surface, or the hypotenuse of the prism, *must* be good to a tenth wave or better so as not to change your instrument's performance. Just *test* your scope *with and without* your diagonal. Many lenses and mirrors have received the blame for misbehavior of their small associate, the star diagonal. The mirror diagonal is recommended for short focal ratios and for high accuracy in larger sizes. As shown at (c) a 90° amici or "roof" prism may also be used. This unit turns the light 90° and completely erects the image, making it fine for terrestrial use. This amici system is often satisfactory for small instruments and lower-power astronomical observing; however, bright objects may show a bright line "spike" from the roof edge, and definition can suffer more or less at high magnifications.

The *solar wedge* or prism as developed by Herschel, shown in the figure at (d), is a modest angled wedge of plain glass without mirror coatings. The optical-quality first surface deflects about 5% of the sun's light to the eyepiece where suitable neutral filters further reduce the light sufficiently for viewing. The major portion of the sun's light and heat passes on out through the back. The rear surface of the wedge differs in angle enough from the front surface to throw its reflected light out of the field of view; accordingly the inside of the entire housing should be well blackened. While essentially for refractors, such solar wedges can be quite satisfactorily used for reflectors if interchangeably mounted in place of the regular Newtonian diagonal. Of course the main mirror may also be left unaluminized for a true solar telescope. At (e) we see how an *unsilvered* pentaprism can be used for solar viewing. The utility of the pentaprism is not fully appreciated in view of its ready availability (war surplus). In the reflector such an *unsilvered* pentaprism, properly mounted in place of the diagonal, can serve for solar observing. Such a pentaprism *with* silvered reflecting surfaces so mounted turns the image right side up for terrestrial viewing with a reflector. The image is still reversed right to left; however, another type in the same position, the surplus penta–amici (roof) prism, yields a fully erected image.

A *Barlow lens,* as shown in the figure at (f), can serve to do more than just magnify the image, as discussed under eyepieces. Where the addition of diagonals, solar wedges, cameras, etc., adds to the mechanical tube length to a point where the eyepiece can't reach the focal plane, a Barlow will optically extend this focal plane by several inches. For example, a Barlow of negative 4-in.-focus set for 2X magnification will extend the focal plane two inches and for higher magnifications even more, as may be calculated by the formulas in Fig. 5.3. The fuss of cutting off some of the tube may thus be avoided. A camera can be mounted on a standard reflector by such a unit supplied by Criterion.

Erectors are necessary for terrestrial viewing with astronomical telescopes. Least expensive is the use of a lens system in a long tube that is inserted in the eyepiece holder of refractors or reflectors, as shown at (g). For retaining the same image size the lens system will be twice its own focal length away from the original image plane and the new eyepiece focal point will be as much again farther away. Varying this distance, or the makeup of the lenses, can give one almost any magnification or reduction desired in image size. Two simple lenses *properly oriented and spaced* can make a good erector for $f/15$ refractors, but a system as shown, with two achromats, steepest curves facing, is superior for reflectors or faster systems. For unit magnification they may be a matched pair, but for added magnification the foci should be adjusted accordingly. For 2X, the lens near the eyepiece should be twice the focal length of the one facing the mirror or objective. These lens erectors often limit the view field. In the figure at (h) is the much more elegant and compact prism-erecting system functioning as illustrated — one essentially as used in binoculars or a prismatic spotting scope. It is highly recommended for refractors where focusing range permits the added optical length. Where not, a Barlow used with it can serve for higher powers. Or better yet, shorten the tube.

Filters rate after appropriate eyepieces in real utility for proper observing. Three basic types are of interest: colored, neutral and reflecting. The function of colored or neutral filters is illustrated at (i), where a red filter transmits red light but internally absorbs blue. Other colors transmit and absorb accordingly. Various uses of colored filters is covered in Chapter 8. Neutral density filters function just the same, but absorb a uniform amount of *all colors* in a percentage according to their density. A filter absorbing 90% and transmitting 10% of incoming light is said to have a density (D) of 1, and one transmitting 1% a density (D) of 2 and so on, the density being merely the logarithm of (100 ÷ per cent of light transmitted). Since such neutral filters actually absorb light and heat internally, they may crack from heat when placed at the eyepiece of larger aperture telescopes, particularly those of high aperture ratio that concentrate the heat in a smaller sun's image. For safety, some other means, as a Herschel wedge, *must* be used to remove 90-95% of the light and heat *first*. Another solution is to use a reflecting filter as shown at (j), one of clear glass coated on the first surface with aluminum or suitable reflecting material to reflect *back out* of the system *most* of the incoming light and heat. If used over the front of a small telescope it can be of optical quality glass; if at the focus of a 6-in. or larger reflector it should by all means be of quartz, since glass ones have cracked. The added cost of quartz in such small sizes is negligible.

Possibly the most useful accessory for solar observation is the simplest one, a white cardboard screen on which the sun's image can be projected and viewed safely by a group. A holder for this can be simply made. Figure 5.5 shows the partial phase of a solar eclipse being projected by the eyepiece onto a large screen for student viewing. A small commercial projection screen is shown in Fig. 5.6 along with two less known but quite useful accessories for demonstration and teaching, the dual observer's instruction eyepiece and the rotating multi-eyepiece turret.

A most interesting but little appreciated eyepiece accessory is the ocular spectroscope, which spreads a star's light out into a colored spectrum (as a glass prism does with sunlight). With 6-in. instruments the variations in the distinctive absorption spectrum lines of the brighter stars and the colored emission lines of glowing bright nebulae open to many a fascinating new area of investigation.

FINDERS, DRIVES, CIRCLES & MECHANICAL

A "finder" is usually a small low-power refracting telescope covering a relatively wide area of the sky, usually 5 to 10°. Most have cross hairs

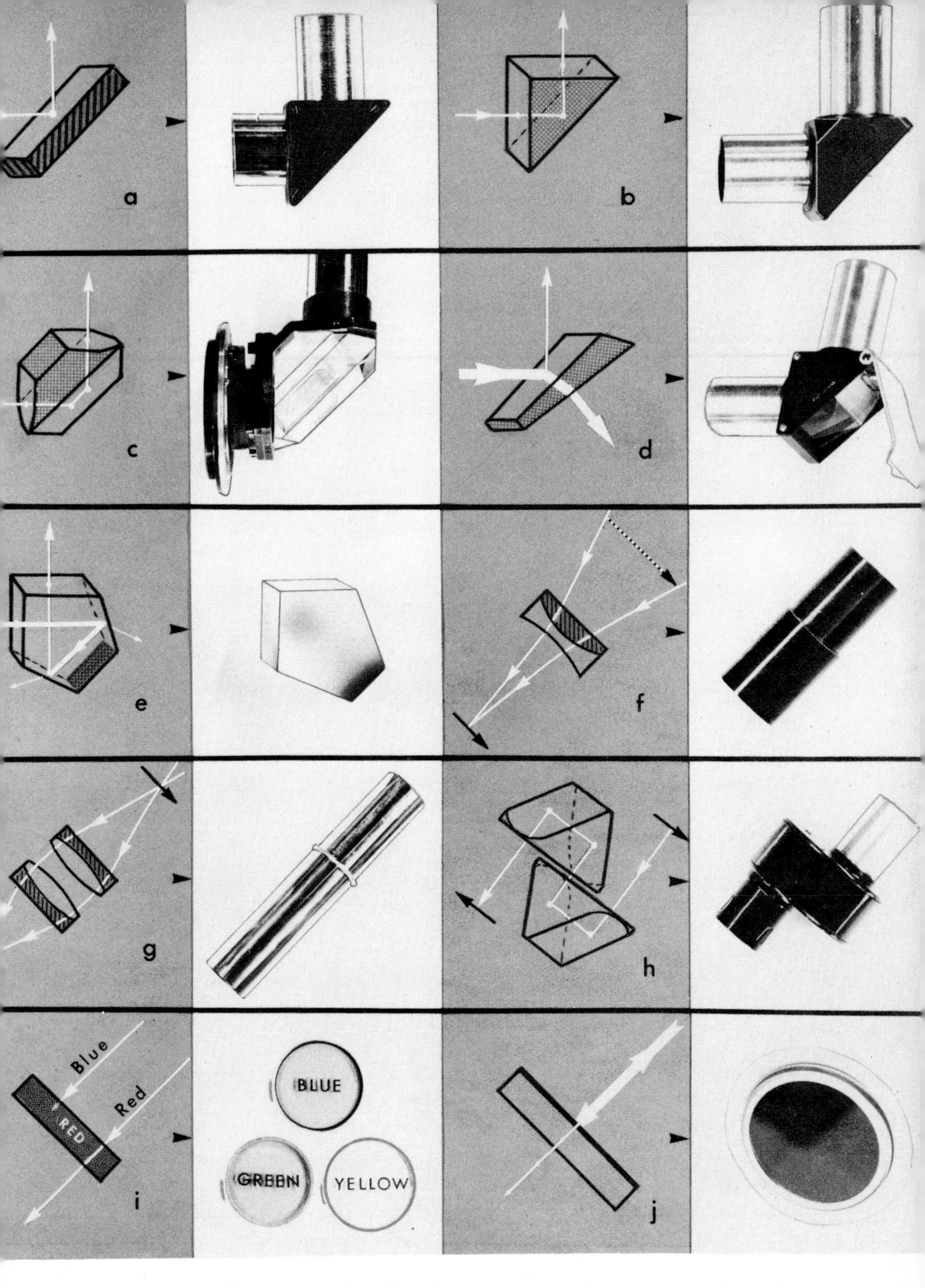

5.4 – **How accessories function.** (a) **Star diagonal – mirror type.** (b) **Star diagonal – 90° prism type.** (c) **Erecting diagonal – amici (roof) prism.** (d) **Solar Herschel wedge.** (e) **Solar diagonal – pentaprism.** (f) **Barlow amplifier – a negative achromat.** (g) **Erecting systems – lens type.** (h) **Erecting system – prism type.** (i) **Color filters – glass.** (j) **Solar filter – reflecting type. Equipment** (a) **courtesy Edmund,** (b, d, and h) **courtesy Unitron. Drawings and photos by the author.**

5.5 – A solar eclipse projected on a screen from the eyepiece of a fine Springfield type 16½-in. reflector, an ideal and safe way for study groups to observe the sun. Photo courtesy Procopius College.

or cross lines centered at focus in the eyepiece view field. One can point the telescope tube in the general direction of an object by sighting along the side if necessary. The object is now centered in the field against the cross hairs, and if the axis of the finder and the mechanical and optical axes of the main telescope are all parallel to each other, we will now have the object snugly in view in our large telescope's eyepiece, at least in a lower-power one. We can then center again and move on up in power. Often overlooked is the sublimely simple but practical gunsight type of finder, *i.e.,* a white inverted V at the front and an M shape or a ¼-in. peep hole at the rear of the tube lined up with the telescope's optical axis, just as with the bore of a gun. A touch of light from your flashlight on the front sight makes object location quick and easy. Enough ordinary finders are shown in the photos of Chapter 4. In Fig. 5.7 are handy right-angle finders for reflectors. They are quite convenient. On a refractor a "Richfield" wide-angle finder (see Fig. 8.6) of *about* 7 to 12X and 50 to 80 mm. $f/4$ or $f/5$ objectives serves a dual function. Several have used 7 x 50 prism *monoculars* in this manner with a cross line glass reticle placed at focus. I adapted an old B. & L. short prismatic 13X spotting scope in a fork slip-on mount to a 5-in. refractor, as shown in Fig. 5.8. Loosening the rack and pinion tube slightly allows you to move it around the tube to be clamped in any convenient position. Shallow fingers in the back prevent its falling off, yet a complete loosening turn of the clamping tube lets these fingers slip by for easy removal. This is fine for viewing dim objects as a sort of bonus "extra" instrument solidly attached and conveniently located. One sees much more with the steady support thus afforded. The low power serves nicely to introduce an object and pave the way for striking high-power views.

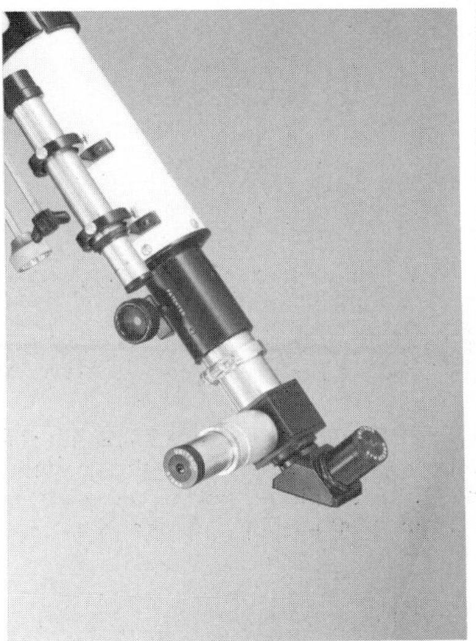

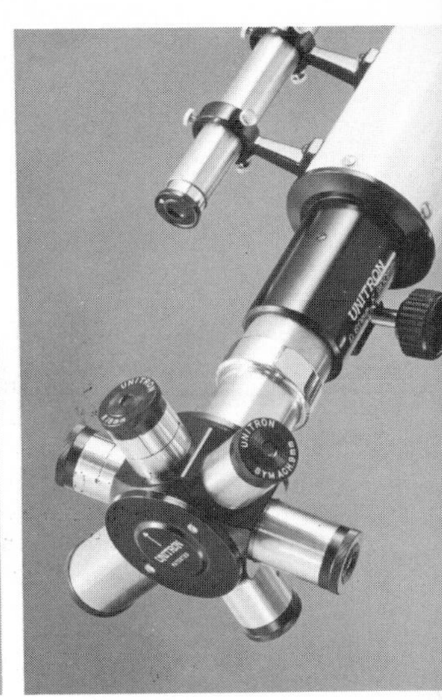

5.6 – **Three handy teaching accessories:** (a) **the solar projection screen system,** (b) **the instructor's dual viewing eyepiece and** (c) **a combined star diagonal – six-eyepiece turret unit. All photos courtesy Unitron.**

For powers above 150 or so, a drive, even of the very *simplest* form, is a real necessity and not just an accessory. Most manufacturers sell adequate drives, often as an added extra, primarily to keep the telescope price down to an unstartling figure, and not because they do not recommend them. For serious study all refractors of 3-in. and up and reflectors 6-in. or over really should have equatorial driven mountings. In Fig. 5.9 we see two well-known drives that can be added, an economical (but adequate) one by Edmund and the refined E. Byers Co. sidereal "slip circle" drive.

Setting circles are not particularly essential for general observing, and I find most owners of scopes with gorgeous precision circles soon ignoring them, since they can locate most objects through the finder once they have "learned their way around" the sky. If one wants to invest in really usable circles (not just decorators) I strongly recommend the type provided by Star-Liner, Criterion, Celestron, Cave and others that function so that once the slip-ring hour-angle circle has been set for a star of known right ascension viewed at the beginning of the evening (I try to remember R.A. of two or three), one can thereafter directly set the telescope to the hour angle of *any* object of interest, swing to its declination, and find it right in the middle of the field. This naturally assumes that your scope's polar axis is targeted on the pole and no one has tripped over the power cord! In Fig. 5.10 is another high-grade sidereal by O. Magnusson (now Cave) who also supplies machine-engraved circles, along with a sidereal circle I quickly added to a conventional commercial scope.

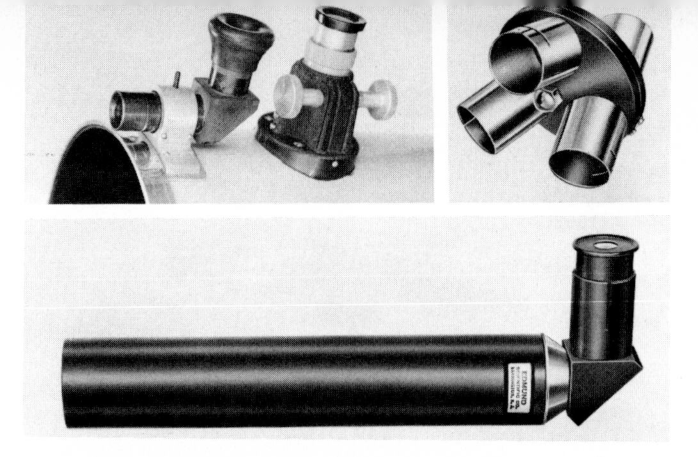

5.7 – (left) **A small Edmund right-angle finder adapted to the author's 6-in. reflector with spring-loaded adjustable mount; very convenient.** (right) **A triple-turret eyepiece holder.** (below) **A fine 7X right-angle finder for reflectors. Equipment and photos by Edmund.**

Certain *mechanical* features are worth considering. Eyepiece focusing motion can be by rack and pinion, spiral thread or simply sliding the eyepiece in and out. Regardless of the type, smooth non-jerky action is a "must." For refractors the rack and pinion is ideal. Those with helical (slantwise) cut teeth are smoothest. A little slack in changing direction is unavoidable, but too much looseness is frustrating in trying to focus. The *f*/8 or faster reflector requires precise short travel focal adjustment that

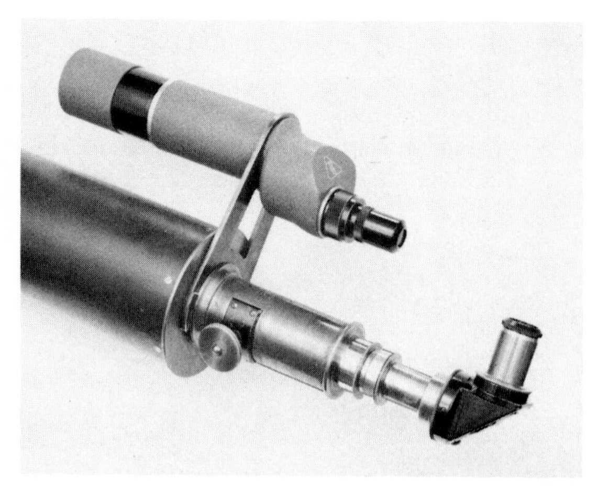

5.8 – **An early model Bausch & Lomb 13 x 60 mm. spotting scope serves here both as a finder and a Richfield viewing scope for a 5-in. Alvan Clark refractor. The Bausch & Lomb current Balscope Zoom (15-60x) wide-angle eyepiece is fantastic for this use. May be swung and clamped at any point about the tube for convenience. A 7 or 10 x 50 monocular could also serve well. Photo by the author.**

5.9 – (left) **An economical but serviceable motor telescope drive by Edmund.** (right) **An excellent sidereal drive unit by E. Byers Co. Photos courtesy the suppliers.**

a coarse rack and pinion does not truly provide. On the other hand a coarse spiral thread works very well. The eyepieces shown in Fig. 5.11 solve this problem neatly, providing *both* rack and pinion and spiral focusing. I've used several of this patented form. The largest size should be of great interest to astrophotographers as would the large unit by Unitron.

Slow motions too must be smooth in action. Without a drive, one in right ascension is most convenient with higher powers. A fine-adjustment screw-in declination certainly is an aid in centering an object in the field. Flexible cables are often used, but the relatively inexpensive electric slow motions with cable and hand control box now available are a real joy to observers and are highly recommended as a useful addition to a fine precision instrument. A. Jaegers has introduced a fine new clock drive.

5.10 – (left) **A photo showing a slip-ring sidereal drive with its precision engraved circle.** (right) **A similar sidereal hour circle has been slip fitted on the worm gear hub and a double pointer attached to the polar axis end of a commercial unit by the author to ease object location considerably. Precision circles now from Cave.**

5.11 – **These patented eyepiece holders combine the merits of both rack and pinion and spiral focusing mechanisms. The giant one at left provides 2-in. clear aperture for extra wide field eyepieces and cameras. Meade also has a very fine series. For refractors contact Unitron, Jaegers, or Edmund. Equipment and photo by Telescoptics.**

Instruments usually are adjustable for latitude, but it's well to check that they remain firmly clamped when once adjusted — many easily slip. The type of head that permits horizontal rotation on the tripod head or pier is much to be preferred, along with an added bubble level, as described in "Helpful Hints." Where rollers are added to support stands, turn-down hand screws should be added for firm positioning at the viewing site.

Tripods and piers deserve a little comment. An absolute requirement is that both be as sturdy as you can obtain, consistent with being able to move any portable type. Sturdy does not necessarily mean heavy. Every instrument quiver magnified over a hundred times is much more noticeable than appreciated. As regards tripods, my second prime requirement for those used with refractors is that they have a geared-head central column regardless of size. It's true a diagonal eyepiece comes first for convenience, but the ability to move the scope up or down 12 to 18 inches is a great bonus. Figure 5.12 shows a fine 5-in. Alvan Clark on a sturdy old Zeiss geared column tripod — quite convenient. No telescope company seems to sell these. For small scopes (up to 60 mm. short focus) the largest amateur Quickset brand for cameras is ample. This same company makes fine all-aluminum professional tripods in three larger sizes that are *very* sturdy for their weight but relatively expensive. Linhof's larger tripods (Kling Photo) are also of the very highest quality but a little heavy and equally expensive. Unfortunately all require some adapter to take the scope head. On the other hand, the wood and brass war surplus tripods ($10-$20) with adjustable cross braces have been economically adapted to many a refractor or reflector head with marked success. Tripods should have a strong safety chain or nylon rope about a third to half the way down to help frustrate "tripod trippers." Rigid bar braces are even better where feasible. I once had a superb refractor irreparably wrecked on the pavement when a dog ran into one of the legs! Actually the large aluminum tube column with three wing feet (often easily removable) supplied by many reflector makers makes a much safer and quite good support. Some buy elevator heads and attach these for refractors. Piers are also available for most instruments but nothing much beats a large pipe filled with concrete, or an even larger reinforced concrete pier, going down below the frost line or bedrock. It should taper out below the ground level, so the ground lifts up and off when freezing. Where *any* vibration exists *nothing* but the telescope should be on or touch this pier.

5.12 — An elevating mechanism on a sturdy tripod greatly adds to the convenience of refracting telescope use. Here an ancient Zeiss tripod with elevator has been coupled to a 5-in. Alvan Clark on a Cleveland mount driven by a Mark III sidereal unit. By the author.

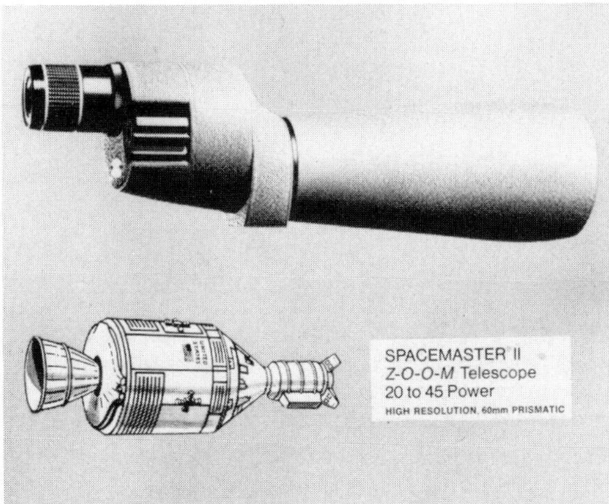

5.12A — This Bushnell Spacemaster Zoom scope with accessory wide-angle eyepiece was aboard the joint USA-USSR Apollo-Soyus space project. Its wide range of powers and viewing fields, coupled with its "prism-folded" compactness made it ideal for all-purpose use in cramped quarters. (See pg. 58.) Photo courtesy Bushnell Optical Co., div. of Bausch & Lomb.

5.12B – **E. Ken Owen proudly takes a close look at the setting circle of his completely equipped 10-in. f/6 reflector. This outstanding example of skilled craftsmanship is a chrome-plated beauty. It's most conveniently mounted on a raised platform with a "flip-off" canopy cover. (See Fig. 3.18.) Owen's complete "Blue Heaven" observatory is pictured in the color section. The equipment and accessories are too numerous and intricate to cover here, however, they are fully described in** Sky and Telescope, **Sept. 1970 and** Modern Astro., **Jan./March 1975. While I've studied many hundreds of amateur-built scopes, this instrument certainly ranks among the most elegant. All hand fabricated —any resemblance to a Cave instrument is purely coincidental. Mr. Owen is also a highly skilled astrophotographer; he works through clear Oklahoma skies. Photo courtesy E. Ken Owen.**

6

BUILD, BUY OR ASSEMBLE?

BUY OR BUILD?

When someone asks me whether he should build or buy a telescope I usually answer with a single word: "Yes." This gets a puzzled expression, which calls for a hasty explanation. Here is the reason I give this answer to those who I think could build a telescope — and most can. When someone suddenly becomes interested in observing, as often follows a first look through an astronomical telescope, it's a good idea to get started right away, before the exciting urge wears off. Many a would-be amateur astronomer struggles through weeks of mirror grinding effort, during which the observing urge slowly fades to be replaced with frustration over a "balky" paraboloid. I therefore recommend one *both* buy and assemble or build a telescope at the same time. Here's how. Acquire at once a *relatively* inexpensive smaller instrument and use it to become acquainted with the skies. You will not only enjoy this, but will know where and how to locate objects, and speculate on how much better the one you build is going to be. Please note that the suggestion was to buy a *small,* not a *cheap,* instrument. A quality 60 mm. spotting scope with a 60X eyepiece (or zoom type) is a good starter — useful in the country too — and as a finder later. Even a 10 or 20X Bausch and Lomb Balscope is a fine "get acquainted with the sky" instrument. So is a pair of carefully tested binoculars. *Binoculars and All-Purpose Telescopes,* published by Amphoto, tells you all about these. A simple 60 mm. to 3-in. refractor is so portable it's always useful, especially in cold weather, even when you have a much larger equatorial instrument. Don't overlook the possibility of acquiring a fine 3- or 4-in. used refractor by early makers. You might just luckily locate one at a low price, and if it is checked by directions supplied in this book it's hard for you to lose. Buy subject to test. Look for such names as Alvan Clark, Bardou, Bausch and Lomb, Brashear, Cooke, Dollond, Fecker, Fitz, Gaertner, Mogey, Ross,

Steinheil and Zeiss. Unless you "know your way around," amateur built reflectors had best be avoided. In Fig. 6.1 are shown three refractors I acquired in this manner, a 4-in. $f/15$, a 3½-in. $f/15$ and a 3-in. $f/10$. Each matched any current commercially made instrument in performance. A fine smaller instrument will keep you interested and really urge you on in your assembly or building job. Constructing a 6- or 8-in. reflector is now well in order. To make one smaller is wasteful of both time and money. In good seeing climates the larger is recommended, even for a "first." In poorer and colder climates an excellent 6-in. has more merit. When you pridefully turn your newly completed large scope to the objects you have been closely studying with your small scope your labors will be fully appreciated. You'll hardly recognize the old objects in their bright new detail, and vast new areas will be opened up for your exploration.

ASSEMBLE YOUR OWN

There are many books on building telescopes; some of the best are listed at the end of this chapter. However, one way of acquiring a very fine instrument that I heartily recommend is seldom mentioned — it's the *assemble-it-yourself-from-what's-available* method. This approach can save you from half to three-fourths of the cost of a completed commercial instrument. It can be immeasurably faster than *completely building* your own — especially if you make your own mirror. Furthermore, it is a most pleasurable way of getting a liberal education in how telescopes go together, and more important yet, how they work. In its simplest form one can acquire a "kit" from such sources as Edmund, Optical b/c Co., Criterion, or the needed parts from such suppliers as American Science Center, Edmund, or Jaegers, etc. Most quality telescope builders also make fine optics. Jaegers even sells excellent long- and short-focus 6-in. (and smaller) refractor objectives at a reasonable cost. Many a refractor has been assembled from these at a fraction of the cost of a commercial 6-in. (See Fig. 6.2 — left).

For those of the younger group with quite limited means who are interested in a "first" small refractor or reflector the book *Telescopes You Can Build* from American Science Center or Edmund is outstandingly good. Its companion, *How to Use Your Telescope,* is a worthy addition. These companies have everything for such assembly, even for larger sized units. Obtain their catalogs.

Adults may wish to shop around more from lists in the journals *Sky and Telescope, Astronomy* or *Modern Astronomy*. It pays to write quite a few places and study their literature on tubes, cells, eyepiece holders, diagonal supports, finders, mountings, tripods, etc. Quite often one can assemble a basic scope first, without drive or other "flossy" accessories, at a price immediately afforded — then add accessories as fortune permits.

6.1 — Three- and four-in. used refractors often become available. These can be excellent buys. Shown are a 4-in. Bardou and 3½-in. and 3-in. Alvan Clarks so acquired. Each is optically perfect. Photo by the author.

There is no restriction as to where to get the parts for a scope. The 4-in. *f*/10 refractor of Fig. 2.4, was made from military surplus optics, an ancient equatorial head circa 1900, an old projection lens rack-and-pinion unit for eyepiece holder, stray tubing and machine shop scrap. It works perfectly. In Fig. 6.2 (right) is a low-power wide-angle 5-in.-aperture Richfield reflector designed to be exactly the right size for hand holding by cradling in the arms. I made the *f*/4 paraboloid, but the eyepiece, micarta and aluminum tubing, etc., were surplus with some machining required.

In Fig. 6.3 may be seen a somewhat more unusual type I once designed and built. The optical path of a 6-in. refractor lens of 90-in. focal length I "acquired" has been optically "folded" with a quartz optical flat at the bottom. The convenient swivel right-angle eyepiece holder is at the top and a knob flips an internal finder in and out of the same eyepiece's field of view. It is a pleasure to use. The last I knew, because of its compactness, this instrument was headed for horseback transport to survey observatory sites on hard-to-reach mountaintops. Here is one excellent way for a group to use an 8- or 10-in. refractor lens that may have been acquired or is "lying around," without building a large observatory. Simply optically "fold" the telescope, using a flat only half an inch larger than half the lens diameter. Even the best are reasonably priced in this size. In Fig. 6.4 we see a multi-purpose terrestrial–spotting–astronomical telescope with three-power rotating eyepiece turret: 25X (wide-angle), 50X and 100X. It's strictly an assembly-building job. The main head is from a discarded large old transit having a fine "silky" cone bearing. The yoke is of bent ⅛-in. aluminum

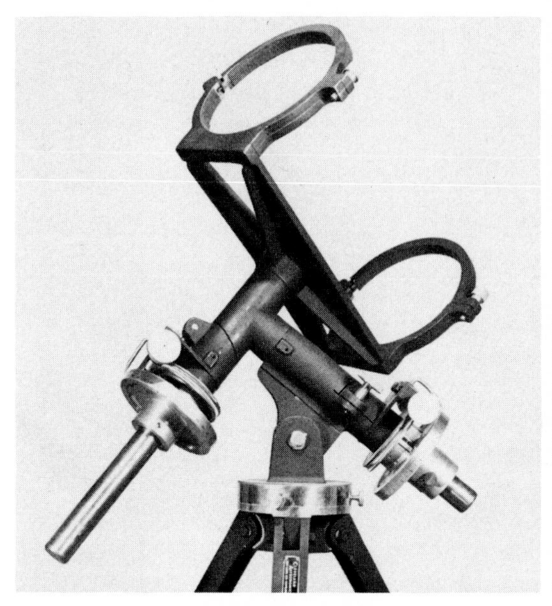

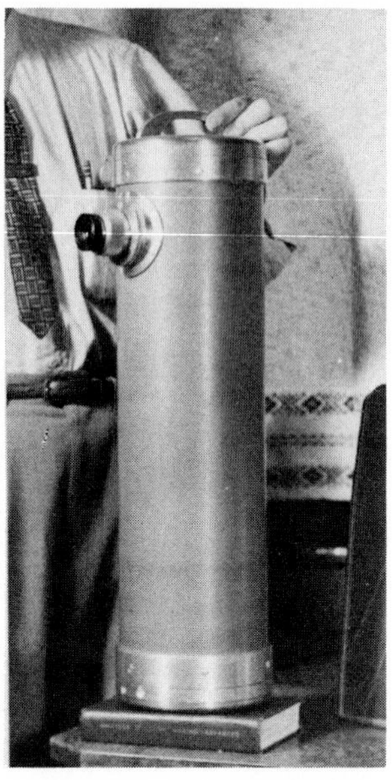

6.2 – (left) **Many mountings may be purchased in parts for home assembly.** (right) **A 5-in. f/4 Richfield reflector designed and built by the author. It's cradled in the arms for steady wide-field viewing and dimmer sky objects. Tuthill sells a closed tube 4-in. "beauty"—an excellent buy.**

with inserted brace and slow motion. The head may be used either as an equatorial for astronomy (left), tilting to one's latitude, or shifted to altazimuth (right) for terrestrial use and target spotting. The image is properly erected by the one amici prism and two-mirror system shown. In this design one can conveniently sight along the tube's top, like a gun. The first mirror slides out to be replaced with a glass solar wedge unit, letting the sun's heat pass down and out through a hole provided. Filters slide under the eyepiece. Objective is a 4¼-in. air-spaced *f*/7.5 super achromat from a submarine periscope. The big eyepiece is Edmund's 1¼-in. No. 5160 (a beauty) "slimmed up" on a lathe. The other two are 16 and 8 mm. surplus orthoscopics. The most "all purpose" instrument I've ever had. I hope these examples show that you can assemble *anything you want to,* since there is endless material available now to work with.

THE BUILDERS

Since *all* the details of building a complete telescope crowds a good-sized book, and has no place in this partial chapter, this space will be devoted to illustrating many fine examples of workmanship to give you ideas for incorporating in your own designs. There are ample books available on mirror making. Edmund's *Homebuilt Telescopes* is excellent.

Since it's natural to show only the best efforts of others, before we come to these I'd like to show three of my earliest very simple telescopes (Fig. 6.5) with no apologies, since in their simplicity they performed quite as well as later more elegant models. Don't let fancy instruments, as fine as they may be, prevent you from building a simple wooden instrument if you want to. It's doubtful if *any* tube performs better than a square one of four boards nailed together. Nor need this advantage be lost in precision instruments. A highly refined version of the square non-metal tube with its proven merits, as built by John P. Wikswo, is shown in Fig. 6.6 (left). At first glance it might fool you by its simple appearance. Actually it's a precision telescope of clean and practical design, with a dual shaft providing a sidereal drive of the "set once for the night" type with the circle handily at the top. Note the totally adjustable head, triple turret, angle finder and elegant substantial base. At the figure's right is one of the neatest *all wood* skeletal reflector tubes I've ever seen of about 10- or 12-in. aperture by Victor Lear of the Rochester, N. Y. Club.

While types and designs of equatorial mountings are unlimited, let's look at those most widely accepted. By far the most used by amateurs, or commercial makers for that matter, is the *German* type shown first in Fig. 6.7. The instrument is a fine example of workmanship by Gene Kada. It's now at the Fieldston School Observatory.

6.3 – A long-focus 6-in. refractor is reduced to an easily manageable instrument by optically folding the light path with a precision mirror. The diagonal eyepiece holder is now near the top of the tube and conveniently swivels about for ease of viewing. By the author.

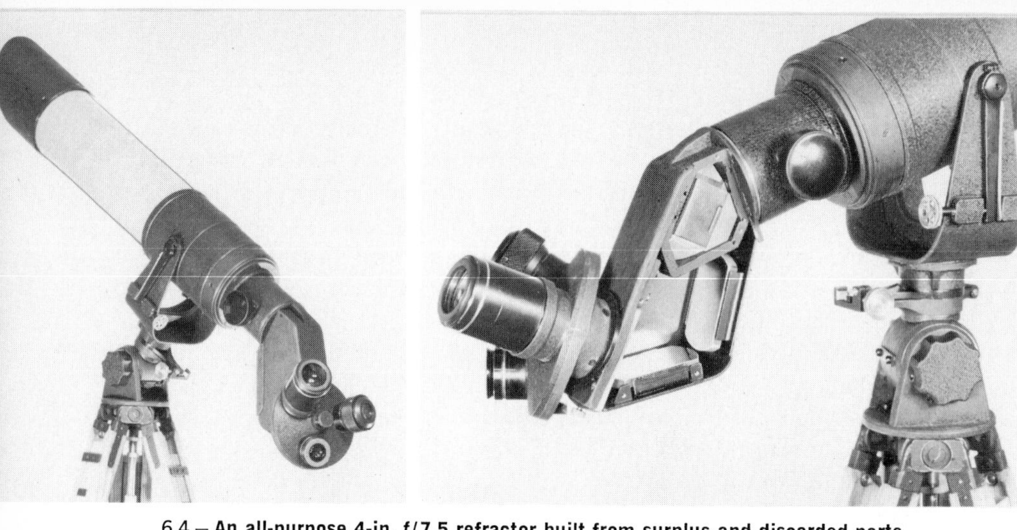

6.4 — **An all-purpose 4-in. *f*/7.5 refractor built from surplus and discarded parts. Turret eyepiece provides 25X Erfle extreme wide-angle and 50 and 100X orthoscopic eyepieces. Filters slide underneath. It's a super target spotter and a good portable astronomical scope. First mirror is replaced by Herschel wedge for solar observation. By the author.**

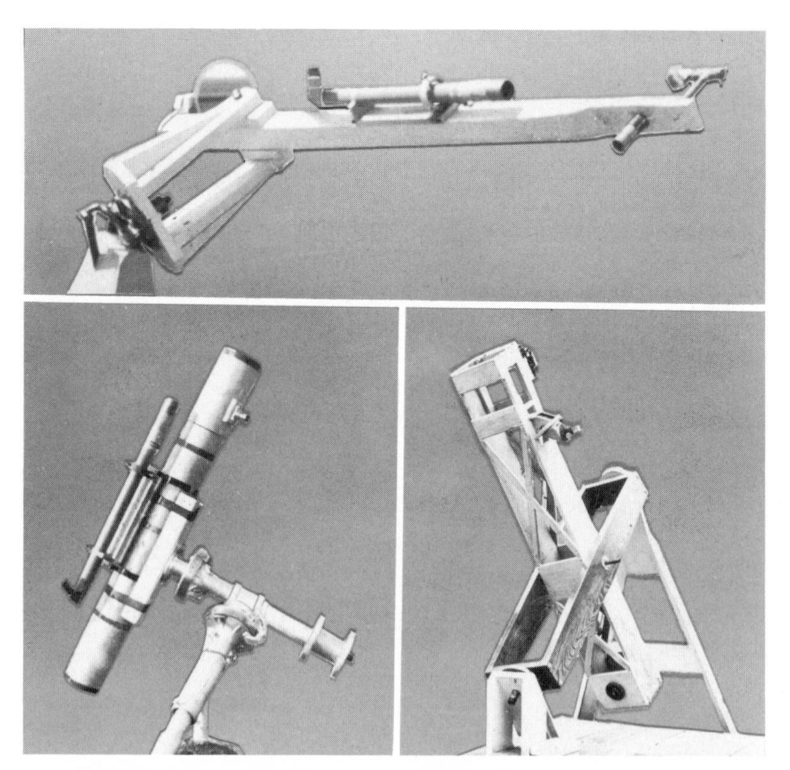

6.5 — **Here are three extremely simple mountings: a wooden spar-yoke 10-in., an auto axle mounted 6-in., and an all-wood English double yoke 10-in. These early efforts by the author actually were superior in optical performance to certain far more mechanically elegant and convenient later jobs.**

Next we see the *Cross-Axis* telescope with a pier at each end of the polar axis. This type is exceptionally sturdy for its weight. This smooth, well-balanced example of the two-pier design is by Cullen G. Scarborough — a telescope that should give future makers some ideas (see *Sky and Telescope,* July 1962).

The *Fork* type shown in the next two photos is increasingly popular, particularly for the newer lens–mirror catadioptric types. It's always been popular for Cassegrainians. The one at the right is a Russell Porter design with a beautiful sturdy hollow fork and the huge upper polar bearing so typical of his sturdy smooth-flowing designs. It has both Cassegrainian and Newtonian heads. The other exceptionally rugged cast aluminum fork mounting illustrated is an outstandingly fine example of overall neat workmanship by William Mason.

The *Springfield mounting* with its fixed eyepiece directed down the polar axis is made with a great many minor modifications, a fine, clean example of which is shown in Fig. 6.8 bottom (maker unknown). Usually the counterweight goes up and over the observer's head as illustrated at the right; however, the weight can be on an opposing underslung apron, just missing the observer's stomach and lap! At the top of the figure we see an exceptionally fine *Pasadena* mounting by G. L. McFarland. This type, invented by F. M. Hicks, differs from the Springfield in basic construction as shown in the photo insert; yet here too the observer sits relaxed in one position, as shown, looking down the polar axis (see *Sky and Telescope,* June 1958). In the last two types the telescope, not the observer, does *all* the moving around from object to object in the sky.

6.6 – (left) **The advantages of the square non-metal tube have been combined here with a superb sidereal drive mounting in this 6-in. reflector by J. Wikswo.** (right) **An elaborate open-lattice all-wooden tube in this neat larger reflector by V. Lear definitely will not cause heat storage or tube current problems. Note appropriate shielding at top and bottom and substantial finder.**

6.7 – **Mounting types. The German form is most widely used. Shown first is an 8-in. reflector on a splendidly proportioned and executed mount built by Gene Kada. Photo by A. L. Copp, courtesy The Sky Publishing Corporation. To its right we see the steady cross axis double pier type, a cleanly designed well-proportioned unit by C. Scarborough. Second in popularity is the yoke mount. At lower left is an unusually husky yoke and mirror cell by William Mason. Next, the clean-flowing lines of a Russell Porter designed sturdy hollow yoke with its extra-large upper-polar axis bearing so essential for stability. Photos by the owners.**

Perhaps the *English double yoke* should be mentioned, although amateurs do not use it widely, possibly because as it is commonly made one cannot view the polar area. It's illustrated in the simple wooden yoke of Fig. 6.5 and imparts great sturdiness to otherwise flimsy wood or iron pipe types. Furthermore, it needs no counterweights. These then are the more common telescope types widely used.

There is no better place to see outstanding examples of fine workmanship than astronomical club meetings or conventions. One of the longest established and best known meeting grounds for amateur telescope makers is at Springfield, Vermont — the "Stellafane" meeting. Let's present some photographs of a few outstanding examples of the many superb amateur-built telescopes displayed there over the years, but first let's learn a little more about "Stellafane."

Many of us knew Russell W. Porter as the man who had much to do with the overall design of the mighty 200-in. Hale telescope at Mt. Palomar,

6.8 – **These are the ideal telescopes for the observer who likes to stay seated and let the telescope do the moving. One looks steadily down the polar axis in either the Pasadena mount, a fine example of which is shown in the upper two photos as built by G. L. McFarland, or the Springfield type, illustrated below by a well built one photographed at Stellafane some years ago. A single counterweight arching overhead balances the entire instrument. See text.**

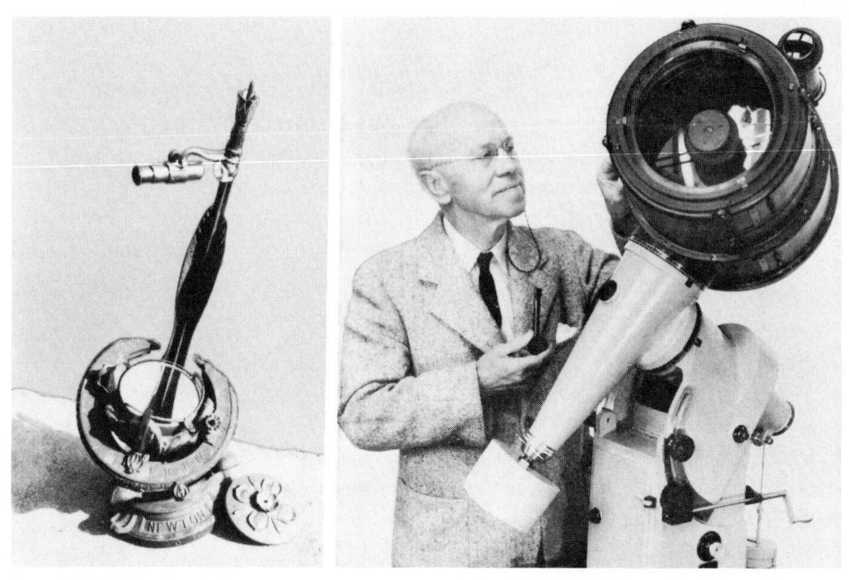

6.9 – (left) **This small Porter Garden Telescope of the early 1920's was a forerunner of the split-ring design of the mighty 200-inch at Palomar.** (right) **Dr. Porter poses beside a high speed *f*/1.0 Schmidt camera so representative of his sturdy, compact designs. Garden scope photo courtesy Bob Cox. Photo of Porter courtesy J. S. Fassero, permission Western Lore Press.**

and for the unsurpassed clear technically faithful "cut away" drawings used in this and many other optical efforts. Fewer had the privilege and pleasure of meeting Dr. Porter at the now famous "Stellafane" summer meetings for amateur telescope makers, jointly held by the Springfield Club of Vermont and the Amateur Telescope Makers of Boston. Over the many years at the Springfield clubhouse up on the crest of nearby Mt. Porter the stars have looked down on some of the world's finest amateur-built telescopes. Here Russell Porter and the Springfield Amateurs, with the assistance of A. G. ("Unk") Ingalls and the *Scientific American* journal, gave the amateur telescope makers in the U. S. a rolling start that continues unabated under the impetus of the three Ingalls edited *Amateur Telescope Making (A.T.M.)* books, and the two fine magazines I often mention.

One unusually beautiful and distinctive small reflector designed by Porter in the early club days, and briefly marketed, was the decorative but functional Garden Telescope in figured bronze, shown in Fig. 6.9 left. The base bears three names: Newton, Kepler and Galileo. Of the some 100 built, only a few remain. Believe it or not, this split ring equatorial mounting, as invented by Russell Porter's creative mind in the early 1920's, became the inspiration and finally the sound basis for the mounting of the world's largest reflector, the mighty 200-in. Hale telescope at Mt. Palomar (see

Modern Astronomy pages 62-65, July-August 1974, for very excellent coverage of The Porter Garden Telescope by Leo J. Scanlon). Dr. Porter is shown by a fine, modern 8-in. *f*/1 high-speed Schmidt photographic camera so characteristic of his sturdy designs. An original Porter sketch (Fig. 6.10) clearly shows the operation of Foucault's famous center of curvature test for a parabolic mirror, and the fine focogram of an *f*/4 parabolic mirror presented shows what the tester sees. Veterans and beginners alike should appreciate this focogram. Shadows of slower ratios, as *f*/8, will accordingly have much less pronounced shadow contrast. The passing of Dr. Porter and Mr. Ingalls and devoted early members has been a real loss, but fortunately this has not diminished the interest in Stellafane or in telescope building.

Let's now take a look at some fine examples of workmanship recently seen at this meeting ground. In Fig. 6.11(a) an excellent example is a very widely made portable Boston Style as made by this group and others. One extra sturdy leg extends parallel to the polar axis and serves as a secure base for the axis bearing. The other two legs adjust for the local latitude and neatly fold up for transportation — an ingenious and practical "portable." Many superb Boston mountings appear at Stellafane. At (b) we see a neat modified form built by Steve Weber of Rochester, N. Y., from a design that sort of "grew" among the Rochester group. This form minimizes

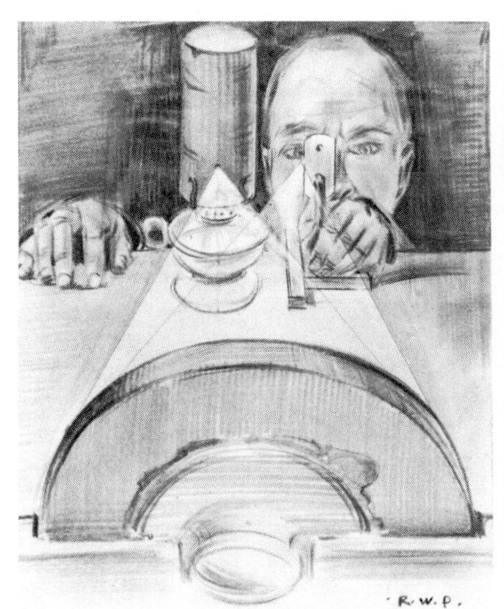

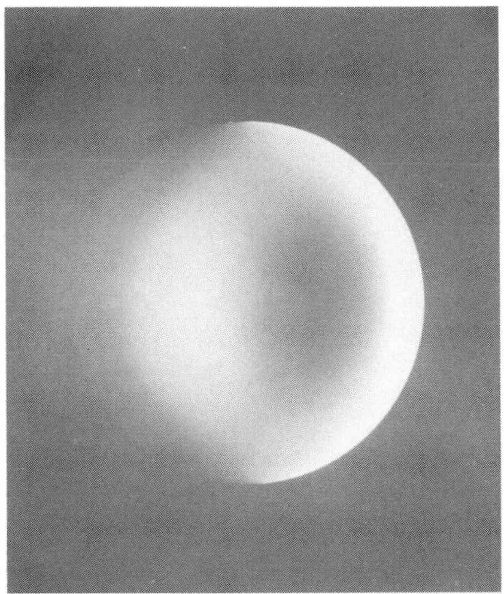

6.10 — (left) **An original Russell Porter sketch depicts the amateur making a Foucault pinhole-knife edge test for a paraboloid at the mirror's center of curvature. Reproduced courtesy Scientific American, Inc.** (right) **The smooth doughnut appearance a happy amateur sees on viewing a perfect short focus parabolic mirror, as photographed by E. McCartney.**

tube interference with the tripod legs and takes maximum advantage of simple pipe fittings plus sturdy large diameter pressure plate bearings as recommended by Alan Gee. At (c) is a clean, neatly built different style that I believe is widely made by the New York City group (and possibly others) under R. S. Luce at the Hayden Planetarium. Note the sturdy construction and fine aluminum castings. At (d) is a slightly varied design, also clean, neat and sturdy, with a head that rotates in azimuth, a practical feature. These are only representative of widely used styles with many variations. It's quite an advantage when a group has exceptionally good design and casting patterns so that all members can benefit. Groups of

6.11 – (a) **The Boston type folding mount with the polar axis housing attached to the main leg, as seen at Stellafane.** (b) **A modified form using pipe fittings and large sturdy pressure plates, built by Steve Weber of the Rochester group.** (c) **A cleanly designed example of the type as made by one group from aluminum castings from a fine pattern.** (d) **A mount that apparently rotates in azimuth, a handy feature. See text. Photos: Weber scope by builder, others from Kodachromes by the author.**

6.12 – Well built amateur telescopes seen at Stellafane. The first two are of the Richfield type for dim objects and wide view fields. Top right instrument demonstrated skilled shop work and unusually fine streamlining. Note the single heavy tube ring support. Each of the instruments shown had high points of merit worth studying. R. S. Luce designed the neat mountings shown far left. See text. Photos by the author.

these have appeared at Stellafane, but each maker soon adds his own "personality" to the rest of the telescope, and often quite handsomely.

In Fig. 6.12 we have first, at the top, two fine examples of short-focus "Richfield" type reflectors, each a well designed unit. Then we come to an exceptional example of good neat workmanship in a pier-mounted instrument. Note the "streamlining" and clean design, rounded mirror cell, and unusually sturdy machined tube rotating mechanism. This was a "beauty." At the lower left is a longer focus portable with good rotating ring and other design features. At lower right is a compact portable of just about the "right" proportions. Note that two of the light cast aluminum feet fold back on the third for transportation. The head is probably rotatable — a fine unit. Certainly these instruments presented should aid in planning a scope just the way you want it to be, most likely different from any shown.

With the advent of the "cats" (catadioptrics, that is) one now sees many of these compact lens–mirror instruments. They make ideally portable closed tube instruments, essentially folding a long focal length into a short tube. Performance in these depends on real care in the making and almost fastidious preciseness as to curve accuracy, lens thickness and concentricity, spacing, and above all, perfect collimation of optical elements. The number made attests to either the wide-spread courage of many, or the contagious enthusiasm of Allan Mackintosh, who for years has been to the "Maksutov Club" what Ingalls was to the A.T.M.'s. Fig. 6.13 presents six of the many well made catadioptrics exhibited. They speak for themselves in their neat compact form. The first fine instrument shown is a neat Maksutov by

6.13 – The now popular catadioptric lens–mirror telescope is well represented by these six, all sturdily and neatly mounted. Maksutovs dominate. As presented, the first is a beauty in blue enamel built by George Keene. The next elegant one was designed and built by Mr. and Mrs. Lucas. Other precision builders, unknown to me, produced the remaining four excellent Maks. Note the beautiful aluminum job, upper right, the sturdy welded yoke at lower left, the exceptionally sturdy German mount at center, and the compact scope and finder-guide scope shown last. From Kodachromes by the author. See text.

George Keene and the second a superb Cassegrain-Maksutov by Mr. and Mrs. Lucas. Unfortunately I cannot associate names with several fine instruments exhibited. Perhaps the owners will point out theirs to their friends and to me for future identification.

A closer look at some of the fine equatorial heads can prove educational. In Fig. 6.14 are nine selected for outstanding merit and clean workmanship. Top row — three practical pipe fitting types from the simplest to a very elegant machined one. In the next row we see a *true* alarm clock drive with unique, handmade, pegged and notched gears. Next we see a white and black beauty of unusually clean and neat design. A superb, refined bronze yoke unit supports the refractor at the right. In the third row, highly polished aluminum gleams on this Boston type mount. The tapered axis housing, German-type mount shown next is outstanding in design and precision workmanship. The last mount is both elegant in overall design and appearance and has a sturdy polar pressure plate disc head, a great stabilizer. I believe these photos speak well for themselves and provide ideas without my further comment.

In Fig. 6.15 are nine of the many examples of unusual instruments displayed, ones that are not only ingenious and unique but were most interesting both to look at and look through. First we see a fine streamlined Brachyt form of "off-axis" telescope silhouetted against the Stellafane Club house. Next, Russell Porter is intrigued by an unusual mounting he's inspecting. We next see an exceptionally sturdy, clean design of the leg type of mount by Mr. R. H. Reniff, who furnished the photo. The instrument's base starts as an extension of the polar axis housing. In the center row we first inspect a prize-winning telescope by D. W. Cogswell, an ingenious, superbly machined and finished type of spherical head that defies adequate description. Aluminum polished to mirror brightness plus a cardinal red anodized tube makes it a real showpiece. Next we have the only western, or cowboy, motif scope I've ever seen, with a saddle-like head, which, with the tube, was decorated in red leather and studs — a pleasing contribution to originality. Next is a fine all-metal mounting of two Porter split-ring equatorial types along the line of the 200-in. design. In the third row is another sturdy all-wood mounting of this type, finished in bright yellow oak with a glossy white tube. The last two, one of metal and the other of wood, nicely illustrate two exceptionally sturdy modifications of this form. In these types the lower end of the polar axis thrusts down into a single bearing. A large disc on the upper end of the axis, usually notched like a horseshoe for the tube to view the pole, rests on edge and rolls on two small widely spaced rollers to complete the sturdy, wide-spread, three-point support — like the roots of a tree. This disc may function as a right ascension circle in some designs. These scopes do not "shake" easily.

6.14 – (top row) **Practical pipe fitting mounts from the simplest to a machine finished job.** (center row) **A true alarm clock–peg gear drive, a black and white refined mount and a streamlined bronze refractor yoke.** (bottom row) **A gleaming aluminum Boston mount, a precision executed German mount and last an artistically designed broad pressure plate mount. Note carrying handle and unusually fine tube saddle. See text. From Kodachromes at Stellafane by the author.**

6.15 – **Some unusual yet well constructed mounts seen at Stellafane.** (top row) **A streamlined Brachyt off-axis scope is silhouetted against the Stellafane club house; Russell Porter studies an unusual mount; and a sturdy leg-type base mount built and photographed by R. H. Reniff.** (center row) **A prize-winning machining job in colorful red and aluminum; a western motif red leather and studded mounting; an all-steel split-ring mount.** (bottom row) **A sturdy golden oak split-ring mount; a steel modified yoke-ring mount; another all-wood split-ring telescope, simple but sturdy. The wood hex tube is very practical. From Kodachromes by the author, except as noted.**

You are now in a fine position to try out your own idea of building, and newer materials can make totally new designs quite practical. From the work presented you can easily see there is *no* limit to what your imagination can produce.

A friend, after reading this chapter, said he liked it, but it didn't tell anyone how to build a single telescope! Quite correct; however, to do this in detail takes a book in itself, as indicated earlier. With this in mind please turn to the following sources for detailed information on building an instrument.

For simple beginner's telescopes refer to Edmund's *Telescopes You Can Build* (75¢). The specific telescope building books one should consider as reference sources are: *How to Make a Telescope* by Jean Texereau (Edmund), *Standard Handbook for Telescope Making* by N. E. Howard (Thomas Crowell Co.), and the widely used *Making Your Own Telescope* by A. J. Thompson (Sky Pub. Corp.). For those who like to dig deeper and design their own, by all means acquire the A.T.M. books (*Amateur Telescope Making* — books One, Two and Three) published by Scientific American, Inc. Books are *very* cheap compared to time and equipment. Why not get several?

All telescope builders and users will thoroughly enjoy a fascinating recent book *Starlight Nights* by Leslie C. Peltier (Harper and Row). The discoverer of a dozen comets, this author relates his many adventures as a star-gazer. His common-sense approach to astronomical instruments and their housing is refreshing and educational. *Nine Planets* (Rev. Ed. 1970) by Alan E. Nourse (Harper & Row) is also fascinating.

A few books of varied nature you might enjoy in pursuing your hobbys of photography, observation and telescope making are: *Russell W. Porter, Patron Saint of Amateur Telescope Makers*, by Berton C. Willard (In preparation for fall 1975 — Band Wheelwright Co., Freeport, Maine); *The Telescope*, by H. E. Neal (Julian Meissner); *John Alfred Brashear*, by Gaul and Erseman (U. of Penn. Press); *The Glass Giant of Palomar,* by D.O. Woodbury (Dodd, Mead, & Co.); *Men, Mirrors and Stars,* by G.E. Pendray (Funk & Wagnalls Co.); *The Telescope Makers*, by Barbara Land (Thomas Crowell Co.); and *Sweepers in the Sky*, by Helen Wright (Macmillan Co.).

7

TESTING AND CARE

MECHANICAL

In telescope testing *both* mechanical and optical checks should be made. In consideration of those who have just acquired an instrument, let's look at the mechanical aspects first. If you have direction sheets, be sure to study them. When "setting up" a telescope, check at once to be sure clutches and clamps are not so tight as to invite damage. After the instrument has been firmly "planted" and leveled, swing the tube about, to check motions of both axes. Whether altazimuth or equatorial, they should be smooth, neither loose nor bindingly stiff. In many forms a *partial* tightening of the clamp smooths the action. In a driven equatorial, backlash should be a minimum, but it's virtually impossible to have none. If the scope is counter-weighted *slightly against* the drive at all times no problem of jerky motion should be encountered. The polar axis should be at least "pointed" towards the celestial pole. Counterweights should slide freely, be tightly clampable, and have some form of safety stop at the end of the declination axis. A dropped counterweight can wreck both the instrument and an observer's foot. Large instruments need counterweights on the tube too, particularly where large finders or cameras are used. Mechanical slow motions should be checked, particularly on the polar axis if there is no electric drive, and so should the declination axis of larger instruments. More important yet, can you reach these from *any* observing position? Often flexible extension cables serve, but electric remote controls are to be preferred for most sizes above 6-inches. Finders should be easily yet precisely adjustable and put into operation by centering an object on the cross hair that has been previously located "the hard way" and centered in the main scope's eyepiece. Adjusting screws should be locked with the means provided. Refined details are covered in the next chapter.

OPTICAL ALIGNMENT

Any optical system *must* be in alignment before valid optical performance tests can be made. This simply means that the optical axis of the mirror, or the correctly put-together objective, coincides with the axis of the eyepiece. This is the most important adjustment for satisfactory observing. The instrument should also be mechanically in alignment — a slightly different and less rigid requirement, unless precise work by circles is planned. Commercial reflectors should be adequately aligned mechanically, but they are rarely in precise optical alignment. For mechanical alignment the declination axis must be exactly at right angles to the polar axis, the tube at right angles to the declination axis and the optical axis of the mirror or lens centered in the tube. A rough mechanical measure and check can get these close enough for *average* work, but if precision is needed, turn to your direction sheet or to one of the telescope making books listed in the last chapter.

Preliminary alignment of the mirror, diagonal, eyepiece train can be made as follows:

1. Remove the eyepiece and center the eye at the rear of the eyepiece holder. Note the diagonal outline itself, ignoring its surface, and see that it's approximately in position and reasonably centered with respect to the axis of the eyepiece tube. At this point backing the eye away allows seeing that the three circles formed by the diagonal, the far end and the near end of the eyepiece tube are uniformly concentric to each other. The diagonal may have to be moved toward or away from the mirror and the spider or eyepiece holder adjusted slightly.

2. Now with the eye centered at the eyepiece tube (a 1¼″ O.D. cardboard disc with a ⅛″ central hole helps) adjust the diagonal so that you see in its surface the main mirror at the bottom of the tube and continue adjustment until the circular edge of the mirror as reflected appears concentric or equally distant from the diagonal's edge.

3. Now look at the reflection of the diagonal itself in the primary mirror and adjust, one at a time, the three mirror-adjusting screws back or forth, by trial and error, until the image of the diagonal is centered in the mirror.

All should now be in fair adjustment. You should see the pupil of your eye in the exact center of a fully uniform concentric series of rings of all items right out to the inner end of the eyepiece tube. It usually takes less time than to read this. Two methods of precise adjustment follow.

To optically align a reflector accurately many textbook methods are available; however, I like best the one worked out by Ralph Dakin and presented with his permission. It can be done any time at any location. No real or artificial *star* is needed, or extra space. True, a little "gadget" is needed in your eyepiece box, but this is simple to make and a pleasure to use. This consists of a tube 6- to 7-in. long of a diameter that slides snugly into the eyepiece holder. It should have a cross hair on the far end and a 1/16″ or so peephole at the eye end. While the gadgeteer will make a fancy one of eyepiece tubing with cross hair reticle and pierced bakelite eye cap, one can be hastily made by rolling and cementing or scotch taping a thick paper tube. Cross hairs (threads) are centered with a ruler and cemented flush to one end and a cardboard disc with a centered peep hole at the other. This should slide smoothly into the eyepiece holder. Now cut a 3/8″ disc or cross of black tape or red auto reflective tape and stick it on the center of the mirror by ruler measure. It can remain there — no harm to performance. A pair of temporary cross threads are centered by measure over the diagonal and scotch taped to the holder's *edge*. Then, assuming the mirror is installed and the diagonal is centrally located in the tube by rough measure, this three-step procedure follows, as described by Ralph Dakin.

1. Insert your special alignment tube, cross hair first, into the eyepiece holder and move the diagonal longitudinally until the cross threads on the diagonal are centered on the cross hair in the eyepiece alignment tube, adjusting spider or eyepiece holder also if necessary.

2. Rotate and tip the diagonal until the reflected spot from the center of the mirror is also centered on the eyepiece cross hair.

3. Adjust the leveling screws of the mirror mount to center the reflected image of the diagonal cross threads on the eyepiece cross hair.

Your telescope should now be in fine adjustment. This method of alignment makes it unnecessary to center the mirror accurately in the tube or adjust the eyepiece adapter so it is exactly at right angles to the main tube. Your telescope will be optically aligned for best performance, although not necessarily mechanically so.

I've found that a flashlight spot pointed down the tube to catch the spot on the mirror can aid in seeing what you are doing day or night. After practice with ordinary reflectors for average use the cross hairs on the diagonal can be dispensed with, as the eye can gauge its center amply close. Systems appreciably faster than *f*/8 with steep cones of light require displacing the diagonal in two directions, one away from the eyepiece and the other towards the mirror to catch the wider part of the cone intercepted. One easy way to ascertain how much is to draw the cone for the field of view used to *full scale* on a long strip of wrapping paper on the floor and *directly measure* the diagonal's correct location and displacement. Dakin states the displacement is only .03-in. for a 6-in. *f*/8, but is .117-in. for a 12-in. *f*/5 and .25-in. for a 12-in. *f*/3 system!

Alignment using a star has its value in two ways. One can roughly align the system as first described without the special cross hair alignment tube, and then turn to a star. Secondly for very highest "final touch" in alignment for high-power work, the star method cannot be beaten, particularly for slight adjustment to correct minor optical shifts from scope handling.

Final alignment by star is made as follows. Locate a bright star, center it in the field and look at it both in and out of focus.

In Fig. 7.1, from the excellent book *How to Make a Telescope* by Texereau, we see how the appearance of a star image, both *out of focus* (top row) and *at focus* (bottom row), differs according to its relation to the optical axis. In each case at the left (A) is what may be seen from a large centering error; *i.e.*, the true optical axis is to the far right in this illustrated case. Your preliminary adjustment should bring you much closer than this. In the middle (B) we see a modest error, and as we come to the true optical axis at the right (C) we have no error, perfect out-of-focus and at-focus images being seen. Of course, at first check the optical axis may lie in any other direction from that shown. Remember too that with a reflector the out-of-focus image will have a central dark zone from the diagonal obstruction. First always start with a medium-power eyepiece and bright star, changing to your most powerful eyepiece and a less bright star for very last "touches." To describe which way to turn the mirror's adjusting screws to correct an error is unnecessary — you'll catch on quickly enough by trial and error and evolve your own "know-how." I like the system where a friend turns the screws while you give directions from the eyepiece. Possibly at times it's less frustrating but slower to do it yourself. One easily remedied bad point on three-screw adjusting systems is that of arriving at one end or the other of screw thread travel. I like to have one of the three adjusting screws locked in *mid position* and the knob removed. The remaining two knobs usually permit ample adjustment. Reversing direction may be necessary, but ends of travel are not reached nor is the focal plane greatly changed. If the locked knob is directly opposite the eyepiece side on the tube, the adjustable two are easily reached.

In refractors having push-pull adjusting screws or other means, examination of the star image and similar trial and error adjustment will end in proper alignment. For refractor alignment in the shop one can take a 4″ piece of eyepiece tubing and turn a 2-in. long wood or bakelite flanged plug to slide into the back. Drill a ⅛″ hole axially all the way through. Now cut the inner end to a *flat* 45° diagonal and blacken its surface. Cement ⅛″ wide strips of reflecting aluminum foil on it to make a cross X — really a coarse *reflecting* cross hair, except where the peephole interrupts at the center. Now cut a ¾″ hole in the side of the tube centering on and *facing* this diagonal (see Fig. 7.4, devices in case). To use: cover the objective with black cloth and slide the unit two inches into the eyepiece

7.1 – (top) The appearance of the out-of-focus ring system of a star (A) in an optical system whose axis is far to the right. At (B) the error has been greatly reduced and at (C) the system is perfectly centered. (bottom) The at-focus diffraction images under the same circumstances as above. See text. Reproduced from Texerau's book, courtesy John Wiley & Sons, Inc.

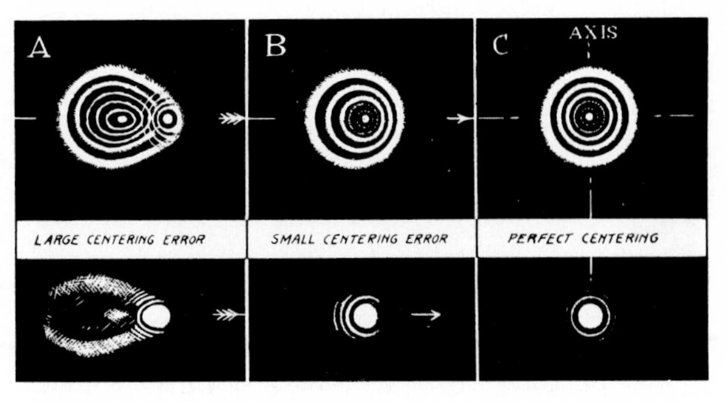

holder. Then flash a spotlight into the opening at the side. Looking through the peephole you will see the bright X reflected from the back of the various objective surfaces — there may be several X's. When they *all* appear centrally superimposed, your objective's axis will be in *perfect alignment* with your eyepiece holder.

STAR TESTING LENSES AND MIRRORS

Nothing beats a real star for checking the *optical perfection* of a lens or mirror, even though in emergency an artificial star in the basement or the reflection of the sun from a distant bright spherical surface can serve quite usefully. Usually a moderately bright star at 45° or so above the horizon *centered* in a medium-powered quality eyepiece serves well. In final tests higher powers and stars of varied brightness will be explored. Star testing must be reserved for nights of "good seeing." On a poor night the stars will pulsate or shimmer and even a telescope known to be perfect will not form a sharp image of any of the stars. Figure 7.2 illustrates at the top the appearance of the bright central spurious disc and first two diffraction rings of a star's image at *extreme magnification* under perfect seeing conditions. The intensity distribution curve below shows the actual comparative intensity of the central disc versus the rings and dark regions between them, as indicated by the height of the curve (also study Fig. 8.1).

I've seen no better coverage of lens testing than that in a now out-of-print book, *The Adjustment and Testing of Telescope Objectives,* by H. Dennis Taylor. Through the courtesy of Sir Howard Grubb, Parsons & Co., Limited, Newcastle upon Tyne, the key plate of this text is reproduced (Fig. 7.3). While these test patterns are refractor based, mirror testers will find this equally useful as soon as the central obstruction effect is observed and recognized in the inside and outside of focus ring pattern. Italicized whole-number figure references with small letters, from here to the close of this section, *refer to this plate*. In *Figs. 10* and *11* we see some of the effects of misalignment as just reviewed, both outside of focus and at focus; hence no further comment is needed.

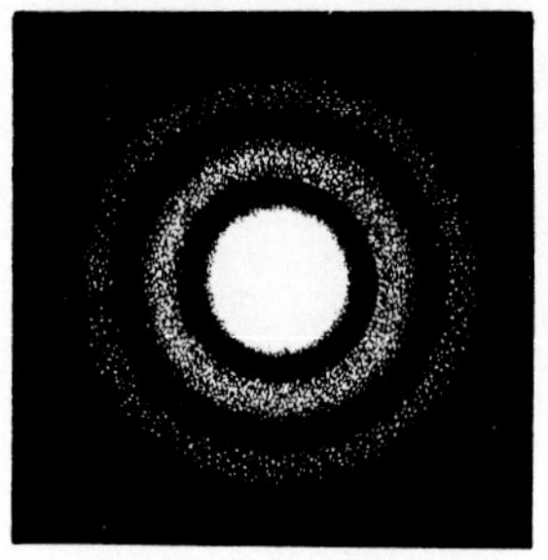

7.2 — (above) **The diffraction pattern of a star at extreme magnification in a perfect optical system.** (below) **The corresponding light intensity–distribution curve for the pattern shown. Reproduced from Texereau's book, courtesy John Wiley & Sons, Inc. See also H. Dall's photos, first figure, next chapter.**

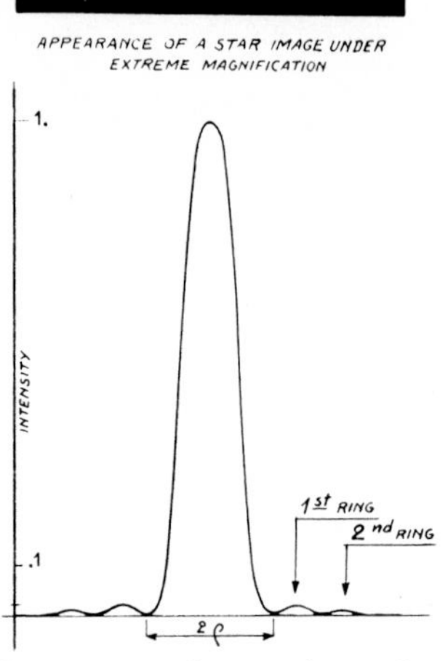

From now on we will assume that our reflector or refractor is perfectly aligned, or as it is sometimes termed, "squared on," so its optical axes pass accurately through the center of the eyepiece. In *Fig. 12* a good image is shown out of focus at *a* and at focus at *a′*. Three point "pinching" in a cell can produce the out-of-focus image *b* and at-focus *b′*. A defect in an objective may product the out-of-focus image *c* — so may a poor diagonal in a reflector. Both refractors or reflectors may have *astigmatism,* with no true sharp focus. As a result of this error images are usually elliptical in shape, reversing at the opposite side of focus, as seen at *d* at one side of the focal plane and *d′* at the other side. If astigmatism is bad, the star will look like a plus sign at best focus. If only moderate, the diffraction disc will be larger than it should be. We must be aware that such astigmatic defects can not only be in a lens, mirror or diagonal, but *in your own eye*. It's easy to check. While viewing the out-of-focus elliptical pattern, notice the direction of its long axis with relation to your eyes. While continuing to observe, rotate your head sort of clockwise and back counter clockwise. If the astigmatic image rotates with the eye, the defect is in the eye. If not, it's in the optical system. Try the other eye. The eyepiece could be at fault but rarely is. *Figures 13* and *14* show such an astigmatic fault inside and outside of focus.

A lens or mirror can exhibit *spherical aberration* to a varied degree; in fact it's a fault common to both. *Figure 15* shows the inside focus pattern caused by positive spherical aberration with feeble central rings and overbright massive outer rings. Subtle changes are lost in reproduction. The complementary appearance (*15a*) occurs outside of focus, the center being brighter than illustrated. Three or four rings out of focus are best for testing. Negative spherical aberration essentially reverses this entire picture. Be *sure* your lens or mirror has reached temperature equilibrium. A rapid change can deform either to produce an apparent spherical error. *Figures 16* and *16a* show how to detect small spherical error by using high power and only two rings inside and outside of focus. Note the intensity differences in the inner rings.

In *Fig. 17* is the spurious disc of a star from a good image under *very high* magnification. Such discs are better illustrated by Horace Dall's photos (Fig. 8.1). *Figure 18* shows what improper three-point pressure, from supports or clamps, can do. Mirrors and objectives are very often clamped too tightly in cells. Retaining rings or clamps should apply *no* pressure.

Figures 19 and *19a* illustrate marked zonal aberration at both sides of focus, a fault best observed 8 to 20 rings outside of focus. This fault is often encountered in amateur-built mirrors and easily spotted by the "null" test described later. Since the zones vary markedly in position and degree, such patterns can vary. *Figures 20* and *20a* show a differing one. In *Figs. 21* and *21a* we have a good optical pattern, except for a region in the center having an extended focus.

In a mirror perfectly corrected for spherical aberration, the appearances of the interference rings *will be essentially alike* inside or outside of focus (if differences of color are neglected in the refractor). In the refractor the outer ring, as seen slightly *inside* of focus (*Fig. 22*), will appear somewhat greater in contrast to the interior rings than when outside of focus, as seen in *Fig. 22a.* A strong green filter designed to eliminate secondary spectrum (as Ilford's "Astra" No. 406) will cancel this effect, and rings should appear identical either side of focus.

If we rack the eyepiece about ¼″ to either side of focus of a star from a perfect objective using a moderately high-power eyepiece, we should see an expanded multi-ring pattern as shown in *Fig. 23*. The reflector will, of course, have its central obstruction pattern in all out-of-focus tests.

Figures 24 and *24a* show what violent mechanical strain can do at two or three points on any optical element. At *24b* we see what bad areas in a glass lens can do, often seen in very ancient instruments but not in modern ones.

Before condemning a mirror or lens, always consider the "seeing" conditions, temperature changes, accessory optics such as diagonals, eyepieces,

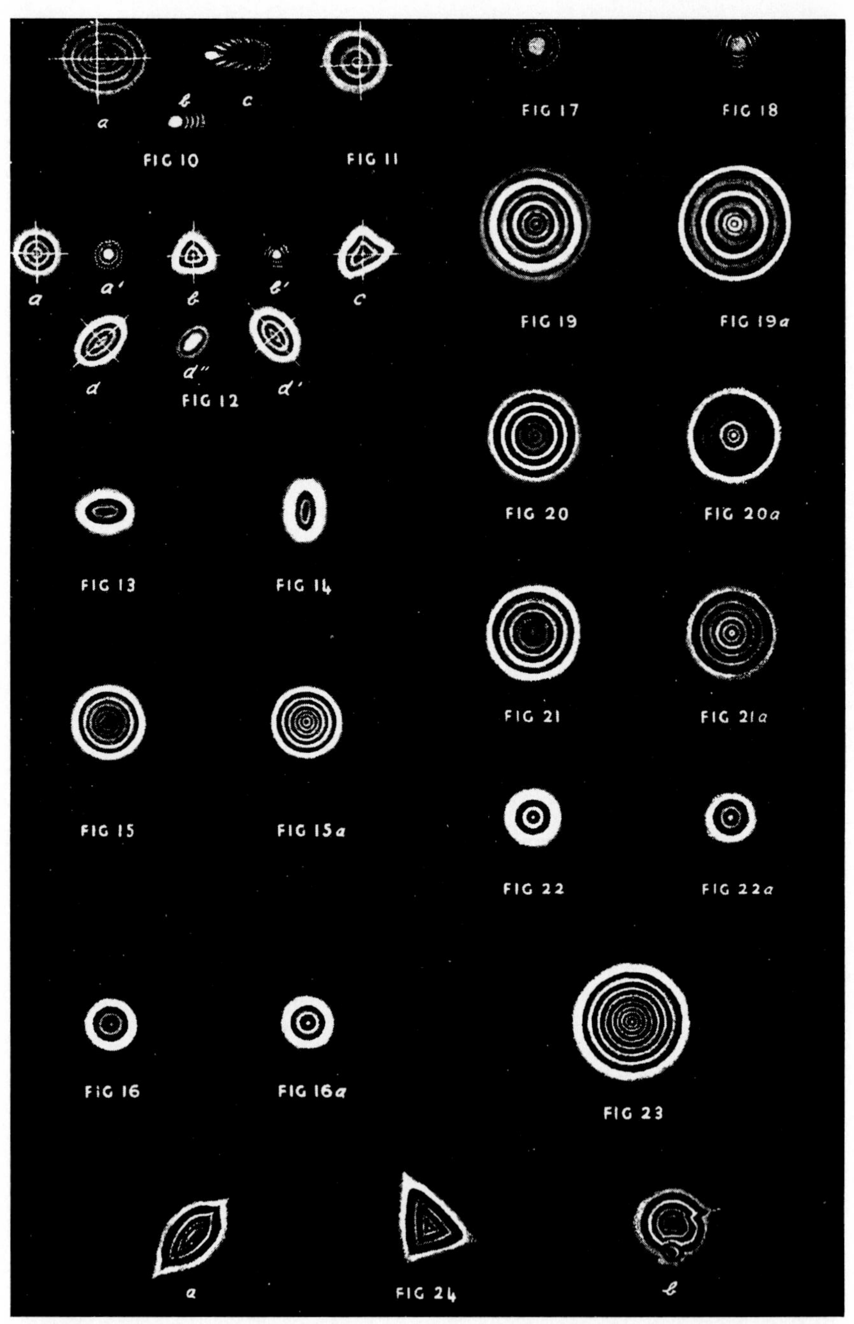

7.3 – **Star diffraction or ring patterns are used here to illustrate and evaluate the more common optical or adjustment errors encountered by the amateur telescope builder or owner. Full details in the text. This valuable plate reproduced through the generosity of Sir Howard Grubb, Parsons & Co., Ltd., Newcastle upon Tyne, England.**

your eyes and eyeglasses. A refractor star diagonal had best be removed before tests, unless you have proven it good. Before condemning a reflector's main mirror for astigmatism, check the out-of-focus image and rotate the mirror a quarter turn in the cell and readjust. If the out-of-focus astigmatic image rotates too, it's the mirror, if not it's the diagonal — not an uncommon but relieving finding. Toss it out and get another. Caution: Be absolutely sure to stay on the *same* side of focus for your out-of-focus test; *i.e.,* either inside or outside, but the *same side* during the test. Otherwise you may throw away a good diagonal or mirror! Check your eye too, as previously described.

Eyepieces can easily be checked for their characteristics by a simple test. Get a 100 line Ronchi grating from Edmund; an 80 to 100 mesh screen will also serve. Hold this quite close to the eye and then look through the eyepiece alone at a bright light source, as a very bright star, or even better a distant light bulb. Shift the eyepiece to and from the eye a bit and observe the line pattern formed by the lines or one plane of the mesh wires. The amount of curvature at the outer edges and the degree of color at the outer field at differing positions can tell you a lot about eyepieces. Try a simple lens, an achromat, a cheap two-lens eyepiece and then an orthoscopic. The latter will be essentially colorless and have quite a flat field (straight lines) for about 40°. It's an interesting test for comparisons.

CARE

Proper care of optical equipment requires a little time and thought but you will enjoy it much more if it's handy, neat and clean. Its usefulness can be extended almost indefinitely.

Eyepieces should have some form of storage case. These can be endless in variety, but Fig. 7.4 shows two conventional types. At the left is a usual fitted box. Any amateur could make such from a surplus chest and a few sawed forms glued in and edged with common green felt or felt from hat brims. At the right is an easier way, a cheap attaché case and two pieces of leftover rubber mattress trim, one with the right sized holes to take the eyepieces. Nothing shifts in this case when closed, and odds and ends fit regardless of shape. Camera cases with foam "cut out" centers are now available and should work well. Eyepieces should be kept as dust free as possible. Bathroom tissue, Q-tips, absorbent cotton and a little detergent and alcohol are good cleaning aids. Always touch surfaces *very* lightly with any solid material and use liquids exceedingly sparingly — just enough to thoroughly dampen but not flood surfaces — except when really washing quite dirty lenses or mirrors with detergents. In this case use distilled (not treated) water for the last rinse. Again it's the "light touch" that's best.

Instruments left outside require careful attention with constant check

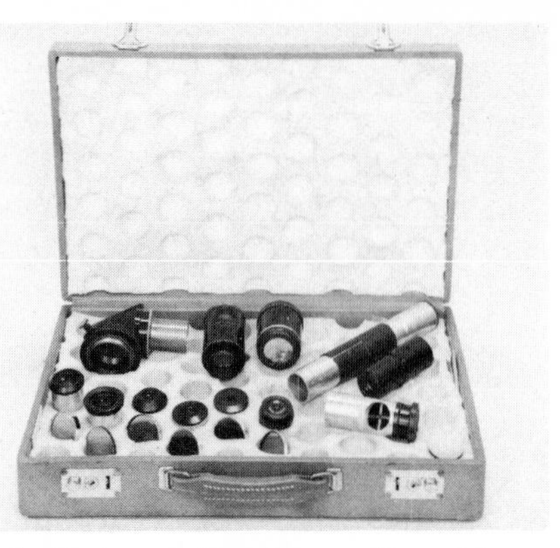

7.4 – (right) **An inexpensive attaché case lined with foam rubber from mattress trim, or one using cut-out plastic foam slabs, is twice as convenient at a third the weight and a tenth the cost or work of the elegant fitted wood case shown at the left. By the author.**

for rust. Rust-Oleum's Damp-Proof Red 769 and 470 aluminum are the best I know for iron under really rugged conditions. Synthetic fast-drying auto enamels or lacquers, if properly applied, have proven to be outstandingly good. Such care as constant oiling and greasing is taken for granted. The silicone cloths used by hunters are fine, but *don't* use them on optically coated surfaces.

Wrong types of covers are often used. An air-tight plastic cover is often tied tightly over an instrument. Moisture inside can't get out, and after the sun heats it up, wreckage of most everything inside soon follows — often in a few hours. A small piece of waterproof plastic can go over the top and part way down the sides, but the overall cover had best be a

7.5 – **A simple cover, as illustrated partially open, made from aluminum roofing on a wood frame that can be slid or tipped off makes fine all-weather instrument protection. The reflector housed was built by John Alfred Brashear, restored by the author, and is now at the Smithsonian Institute.**

treated heavy canvas tarp, which, although water resistant, "breathes" enough to serve properly. It's better if there is some open area at the bottom. Speaking of moisture, taking an instrument in out of the cold presents real problems. A cotton padded or multi-layered blotting paper disk, previously dried in the oven, dropped over (but not touching) either a lens or mirror will dry it quickly. If the mirror can't be reached, tight plastic covers over each end of the tube *before* you bring it in and left on until the instrument has warmed up may help — but then remove them for full drying. Dew caps twice as long as the tube's diameter are a help against dew, stray light, and body heat.

Housing for instruments take many forms. To fit just the telescope, a small enclosure — one that can be slid, lifted, rolled or tipped off — works fine. A 12x20 ft. observatory built from a prefab Sears garage (See my book *Outer Space Photography.*) amply houses two instruments. (Also see clamshell cover by E. Ken Owen, p. 54.)

OBSERVATORIES

Since my last edition, many hundreds of amateur observatories of countless designs have been built, with dozens of plans published in interim journals such as *Sky and Telescope*. Most writers tell how the observatories are built and point out their advantages. Quite naturally, failures are seldom acknowledged. An important question is, "What would they do differently if they built a second one?"

Above all, avoid mass (weight) like the plague. These so-called "heat sinks" include large (12-in. plus), extra heavy scopes. Only a lucky few enjoy steady-temperature areas; most areas experience falling evening temperatures, from moderate to severe. Concrete or cinder-block walls and extended concrete floor aprons are definitely out. Simple observatories of marine plywood (¼″ braced, or ⅜″) of the folding or "roll off" type are very hard to beat. Folding up or down walls make fine "wind breaks." Good lumber for frames presents a problem, since straight-grained, knot-free redwood or hard pine is hard to get and expensive. Consider cutting 1″ marine plywood into 1″x3″ or 1″x4″ strips and cementing two or three strips together using contact cement or Elmer's Glue-all, plus galvanized or aluminum nails. For a structure of this sort, it is best to have fire insurance.

Seriously consider "roll offs" operating at ground level. On steel or wood frames, aluminum siding is good and a fiberglass (white) or aluminum roof is fine. My "roll off" (see Fig. 44 in my book *Outer Space Photography*) was mechanically perfect — beautiful but not suited to the climate. It took until 2:00 A.M to cool off. This could have been avoided by running aluminum walls to the ground and rolling off at ground level — *no* concrete aprons. My Outer Space Observatory II, over the two-car garage in back of my house, is made of fiberglass and aluminum. (Wood and large lawns surround the area — see color section.) It has no heat problems and opens or closes in approximately 10 to 20 seconds. I've visited several large home-built observatories that were so cumbersome and heavy that the owner dreaded taking the time and effort to open them up — even on a good night.

Regarding color, a white roof of fiberglass or aluminum is fine. If you are near the sea coast, protect the aluminum (after acid etching) with white marine gloss paint. Although white is thermally better for the walls, any light color will work sufficiently well.

What surrounds your observatory is also extremely important. Grass right up to the walls and as far out as possible is the best. Did you ever notice that scopes out on a big lawn work fine? Keep asphalting and hot car engines as far away as possible. Try to locate as far as you can from mercury and other lights, or use long light shield-dew caps as a last resort.

8

HELPFUL HINTS

IMPROVING REFLECTORS AND REFRACTORS

In Chapter 4 we stated the faults of reflectors and refractors as commonly made. Most books carry these same complaints, but few offer suggestions for improvement. Let's look at the basic causes for loss of performance.

In the *reflector* the primary problem hinges around the obstruction and scattering of incoming light by the diagonal and its supporting arms. It is *not the amount* of light lost that is of primary concern, since this is a relatively small part of the total light, but it is the *diffraction or scatter* of light by the edges of these obstructions that is truly serious. (This area is well illustrated and covered by Dr. Edward Everhart in the *Astronomical Journal,* Volume 64, pages 455-63, Dec. 1959.)

Increasing the central obstruction above a tenth the diameter (only 1% of the area!) in a reflector noticeably changes the perfect stellar diffraction pattern, and image contrast of extended areas is reduced with startling rapidity with increased obstruction. A 1½″ diameter diagonal used with a 6″ mirror (one-fourth the aperture) obstructs only 6% of the incoming light, yet there is considerable loss of image contrast. The diagonal support edges add appreciable insult to the injury. While such change may be easily noted visually when *direct comparison* can be made under good seeing conditions, it is difficult to show photographically. Horace Dall has been able — through considerable effort (Fig. 8.1) — to show some of these effects (reproduction losses make it even more difficult to illustrate photographically changes in light distribution). Using larger central obstructions helps to illustrate the problem. In this figure we see (left) the effect of increasing the central obstruction going from zero obstruction at the 12 o'clock position clockwise through ⅙, ¼, ⅓, ½, up to ⅝ of the aperture obstructed at the 10 o'clock position. The central pattern is also

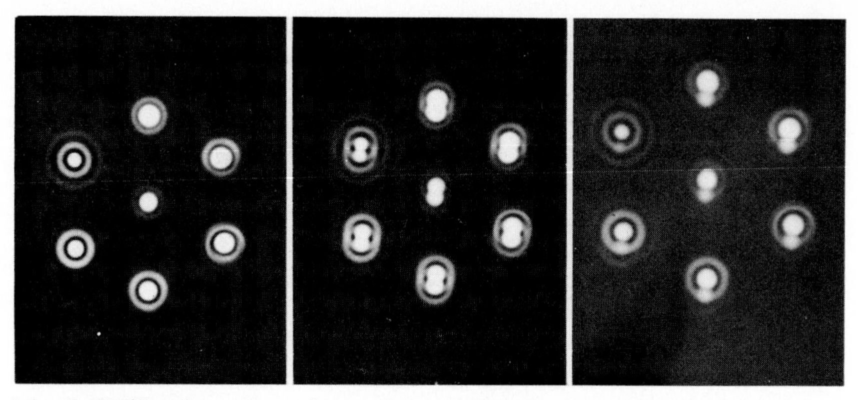

8.1 – (left) **Diffraction pattern of star with a refractor having increasing central obstruction, clockwise from 12 o'clock.** (center) **Equal doubles separated by Dawes' Limit with same increasing obstruction.** (right) **Same conditions but 1.2 times Dawes' Limit separated companion 2.5 magnitudes fainter. Central object in all photos is one magnitude fainter. See text. Courtesy Horace Dall, Luton, England.**

from an *unobstructed* aperture, but from a star *one magnitude fainter.* If we refer to the light distribution pattern shown in Fig. 7.2, what is actually happening is that more and more light is going to the outer diffraction rings as the obstruction gets larger, thus reducing surface contrast of any object viewed. Also note that as obstruction increases the central disc actually gets smaller; this being the reason obstructed apertures can under favorable conditions separate closer stars, yet can fail on faint surface detail differences, as the second photo of the figure illustrates with equal magnitude double stars at Dawes' limit separation. Doubles having low-intensity companions falling on an enhanced diffraction ring may be missed entirely, as illustrated in the third photograph with unequal double stars at 1.2 times Dawes' limit separation — companion 2.5 magnitudes fainter.

What can be done? Easiest first thing is to reduce the diameter of the diagonal as much as feasible. Longer focal ratios (*f*/10 to *f*/15 mirrors) would permit smaller diagonals, but this loses both compactness and illumination. It's best to retain an *f*/7 to *f*/8 standard mirror and use as small a diagonal as possible by means of placing the eyepiece close to the tube and fully illuminating only the central ¼″ or so of the view field, with some light loss (vignetting) over the rest of the field. This would permit use of a diagonal as small as ¾″, an eighth the diameter of a 6″ mirror/7″ tube combination, *i.e.,* a diagonal *one-half* the size of that often used, with reduction of central obstruction effects to insignificant amounts. Such an arrangement is perfect for planetary observation where the planet is centered in the field, although the fall-off of illumination at the edges for other viewing will scarcely be noticed. Be sure the diagonal is well centered and has *no* "turned edge." The focal plane should be within about ½″ from the side of the tube, using the shortest possible eyepiece holder. The very best type can be built along the line illustrated in Fig. 8.2. Here a stalk supporting the diagonal is attached to the eyepiece holder and focusing is accomplished by sliding the entire unit toward or away from the mirror. *The eyepiece to diagonal distance remains fixed.* To calculate diagonal mirror axis diameter for yourself, establish the distance from the center of the tube to where you want the focal plane to be and divide this value by the *f* value number of your system. This gives you the *absolute minimum* diagonal diameter with only the center of view field fully illuminated; less would act the same as cutting your mirror size. Now add to this value the value for the diameter of field you wish fully illuminated. For a 6″ *f*/8 scope with focal plane 4″ from tube center, using a ¼″ fully illuminated field (a planetary scope), we get a ¾″ diagonal; for ½″ full field (general purpose) we have a 1″ diagonal. Note: Such small diagonals are *not for photography* where a larger diagonal, even 1¾″ for our 6″ mirror, is required for complete illumination. The 1¼″ diagonal commonly used with 6″ mirrors is a sort of overall compromise.

Now to *spider supports*. These cause diffraction effects by their edges, as may be noted in photographs exhibiting four or six spikes of light on bright stars arising from the four (Fig. 8.3) or three diagonal support arms respectively. The largest of observatories face this same problem; note four diffraction spikes on the bright star by Pluto, Fig. 1.10. Several forms of support arms are shown in upper Fig. 8.2. The curved ones shown, having radii equal to or one-half that of the mirror, can modify diffraction effects nicely for photography to eliminate "spikes," but do little to improve image contrast for viewing. In fact these special curved forms have been jokingly called "diffraction spreaders" — probably appropriately so. Where an arm support is used for observing, I strongly favor a *single* stiff flat stalk support, say ⅛″ or 3⁄16″ x 1″ or so edgewise to the light, as shown first in the figure and as illustrated by Mr. Simpson's fine workmanship. Many others have reached this same conclusion. This also permits the sliding focus unit as is illustrated. Dr. Everhart prefers the eight-piano-wire support shown last in the drawing.

Possibly a better solution is to use an optically flat plano parallel glass plate to support the minimum sized diagonal via a ½″ hole drilled in the center, thus eliminating spider effect entirely. The plate should be antireflection coated, equally thick to about 1/1000th inch, and with surfaces preferably as good as the mirror in order not to introduce errors. One can obtain these flat parallel glass plates from manufacturers of precision optics. You can test a plate from any source by holding it in front of your conventional reflector and choosing a piece or area that has no effect on an appropriate star image being viewed. Figure 8.4 shows such a diagonal

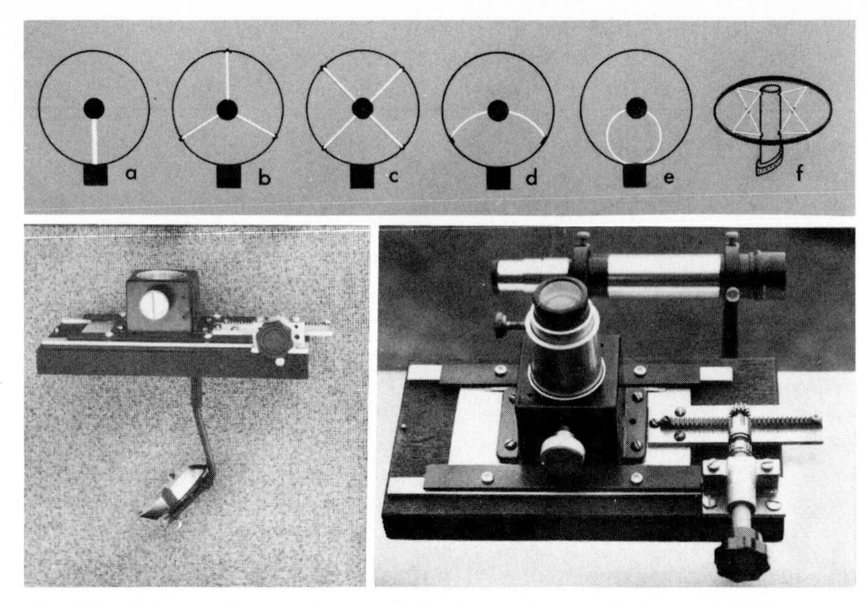

8.2 – (above) **Some common forms of diagonal supports or spiders; one to four straight supports, curved segment and wire supports.** (below) **Two views of a superbly constructed single-stalk horizontal focusing eyepiece holder unit having many advantages. Built by Clark Simpson. Photo by Bob Snodgrass. Drawing by the author.**

supporting plate unit in operation. It is said one can thus have the advantage of the closed tube refractor — possibly on the basis of "what's good for the refractor is good for the reflector." This has turned out to be a "snare and a delusion" with 6″ and larger $f/6$ to $f/8$ instruments with either metal or plastic tubes under my seeing conditions (bad temperature drop). Accordingly a 3⁄16″ to 1⁄4″ ventilating slot is provided all around the plate using three point supports and an equivalent area or more in the mirror cell. This works better.

Another approach is to couple the smallest possible diagonal with a suitable high-quality negative (Barlow) 2X lens located at the tube's wall, to convert the $f/8$ cone of light to an $f/15$ system at the eyepiece. This permits the use of a standard focusing eyepiece holder, the eyepiece further out and longer focus eyepieces for eyeglass wearers! I like the Barlow system and have an 8″ *planetary* reflector in fiberglass tube with a 1.1″ diagonal on an optically flat plate (by Ferson) similar to that just shown. It performs beautifully indeed. A positive 1:1 or more projection (erecting) lens system at this same location permits even a smaller diagonal, say 1⁄2″, but it lengthens the tube and puts the eyepiece quite far out from the tube. A pair of small objective lenses with most curved surfaces almost touching makes a good 1:1 system, or if focus of one nearest the eyepiece is twice that nearest the mirror we shift from an $f/8$ to an $f/15$ system where this is desired. This also erects the image.

The *tube* of a reflector deserves careful consideration. The common heavy aluminum tube (7″ diameter for 6″ mirror) is probably the worst form. If it must be used, it should be lined with ⅛″ sheet cork cemented in place (from your auto supply). Fiberglass tubes are preferred by a great many. Larger instruments can profit by so-called open lattice tubes having closed portions of about the diameter length both at top and bottom, along with a body heat shield near the operator. The best performing tube I ever had was my first when I could afford only a cheap, clumsy, square 8″ (I.D.) wooden tube with a 6″ mirror in a ventilated wood cell — one of those rare advantages of being poor. While mass has an advantage for stability, where falling evening temperature occurs the heat storage can be deadly, steadily coming out again to ruin definition. I once had an old one-ton cast-iron mounted Brashear with a perfect 8-in. mirror (plate glass) in heavy steel tube. It only performed perfectly *once* in two years, an evening of *no* temperature change. It's now in the Smithsonian Institute. Conversely a friend has a 10″ perfect Pyrex mirror in a very open skeletal cell in a 12″ thin-walled (about 1/32″) rolled and riveted tempered aluminum tube with reinforcing rings mounted on an oak cradle. In about 15 minutes or so it "settles down" to work beautifully. Sturdiness is best gained by structural design instead of weight; by large hollow aluminum axes, large flat pressure plates, tripods with a large spread where the legs fasten, and appropriate and skillful light tempered aluminum tubular or angle bracing or reinforcement. If cradle is of metal, insulate it from the tube with cork. Light-weight plastic mirror cells are now available from Optica b/c Co. and others. W. R. Parks supplies elegant fiberglass tubes.

What about ordinary "off-axis" systems to avoid central obstruction? The *faintest* residual image error *not centered* is quite annoying to the eye; hence, I'm not in favor of such, except a rare system or two that can be made only by a most skilled and dedicated worker.

The *refractor's* greatest weakness is the residual color error of its "almost" achromatic lens, producing a colored halo about bright objects and scattering some secondary spectrum light across planetary images to reduce contrast and fine detail. This color error is not really serious in 3- to 5-in. refractors but becomes noticeable in 6-in. (particularly the 6-in. $f/10$ type) and larger sizes. To be more specific, in order to keep secondary color error "reasonable" in ordinary achromats the focal length of a refractor should be three times the square of the lens diameter. Calculations provide these focal lengths and f values: 3″ = 27″ ($f/9$); 4″ = 48″ ($f/12$); 5″ = 75″ ($f/15$); 6″ = 9 ft. ($f/18$); 8″ = 16 ft. ($f/24$) and 12″ = 36 ft. ($f/36$). Hence a 5-in. refractor is fine, a 6-in. good, and with larger refractors one just tolerates color. I once took a reflector owning friend to look through a fine 20-in. observatory refractor. The usual crowd was gathered. To my startled amusement, but possibly not that of the observatory director, he suddenly exclaimed loudly, "What's that terriffic purple halo doing around Jupiter!" One either puts up with this, tries filters, or turns to a reflector.

Some of this residual color can be removed with filters if you do not mind a slightly yellow or greenish object; in fact, it's amazing how soon the eye "forgets" a change in hue. Of all filters I've tested I like best Eastman's photo filter No. 3 light yellow or the glass Leitz GR (light green-yellow) Leica filter. Your X-1 green camera pan film correcting filter is slightly darker but usable. The latter two increasingly remove some of the red error as well as the purple haze. Others may wish to try Eastman's darker yellow No. 4. Obtain an Edmund No. 40,272 diffraction grating for 25¢ and hold it so a single light bulb viewed forms a nice bright continuous spectrum from purple to deep red. You can now quickly check filters to see what portion is removed by rapidly flipping one in and out between your eye and the grating. The filters recommended remove much of the secondary spectrum, yet retain "near" blues and reds. Those testing objectives for spherical aberration, etc., should be aware of Ilford's dark green "astra" No. 406 designed to completely eliminate secondary spectrum of refractors. This it does, but it's too dark for usual observing use. It's true there are also elegant three-lens apochromatic refractors presenting almost colorless images — wonderful instruments indeed when in perfect adjustment. However, they are primarily in a class for millionaires and ultra-fastidious individuals. For the same money a larger optically perfect refractor or a fine reflector is a better buy.

I like a refractor for lunar and planetary work, although I have to admit that an *optically perfect* reflector about 20% larger with reduced central obstruction, as described, can do every bit as well and possibly even better on steady temperature nights. Most of the early comments on the marked superiority of refractors over reflectors was based on comparison between fine commercially made refractors against mediocre (or worse) amateur mirrors. Both amateurs and professionals now can and do turn out superb mirror optics.

NULL TESTING MIRRORS

While excellent parabolic mirrors are made by the regular Foucault Test using zone masks, etc., most experienced amateurs have seen a great many poorly performing mirrors *thought* to be paraboloids made by this procedure. It does not take much deviation from a properly shaped, truly smooth "doughnut" shadow, a small zone, or a bit of turned edge to shift a mirror right down to a poor performer. I'm one of the group who believes in the null test methods; one whereby a uniform graying over of the mirror at the focal "cut-off" point is used as a measure of perfection. The appear-

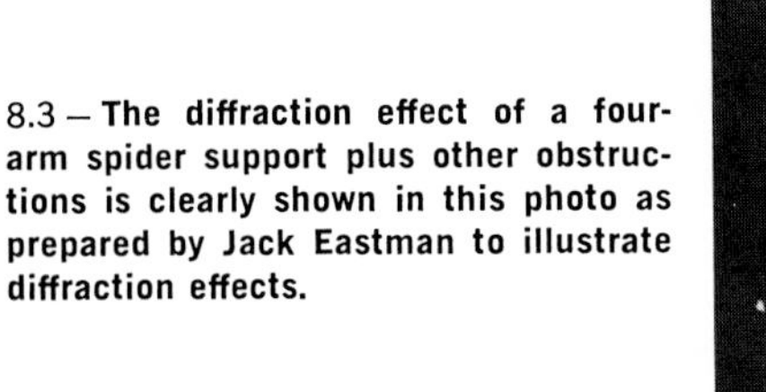

8.3 – **The diffraction effect of a four-arm spider support plus other obstructions is clearly shown in this photo as prepared by Jack Eastman to illustrate diffraction effects.**

ance is that of a perfect sphere as tested with source and knife edge at its radius of curvature. The eye is *very* sensitive to small changes of gray shading, making such a test delicate indeed.

The classic mirror null test, other than using a star itself, calls for a high precision optical flat mirror to act as a collimator to provide parallel light, as though the source were at infinity. Such a flat is far from easily obtainable. Then there is the Dall Null Test as described in A.T.M. Book Three, a note in *Sky and Telescope* by S. J. Warkoczewski, pages 512-15, Oct. 1955, and carefully detailed with equipment and construction by J. Schlauch and R. E. Cox in *Sky and Telescope,* pages 222-26, Feb. 1959. This test is more practical, especially for larger sizes, but still requires a quality lens of known characteristics and much care in its positioning.

Let's see how we can make a fine mirror by null testing requiring no added high-precision equipment. Limitations are few, but specific.

1. A reasonable shop length, as 40-60 ft.

8.4 – **A glass plate, plano-parallel and coated, supports the diagonal for this 6-in. reflector to eliminate damaging spider effects. Used for wide-field observation and photo work. Smaller diagonals similarly supported excel for planetary work. See text. By the author.**

2. Staying with modest mirror apertures and f values — as the *commonly made* 4-in. to 8-in. mirror apertures of $f/8$ ratio. It's *not* for short-focus large-aperture jobs.

Fine mirrors can be more easily made and should be much better "first" mirrors than those made by the conventional Foucault Test. The procedure first came to my attention when presented by Charles Spoelhof under the title "A Simple Null Test for a Parabolic Mirror" appearing in the "Proceedings, 14th Annual Convention of the Astronomical League," pages 101-7, Sept. 3-5, 1960. The basic thinking and practice by Spoelhof is condensed here as follows.

Star or autocollimator tests are based on a light source at infinity to provide parallel light. However, infinity is not an established distance; one usually "approaches" it. To some, infinity may be out beyond an outermost galaxy; but to a youngster it may be just over that hill. Practically, for telescope mirror making either point could serve. What Spoelhof has done can be simply put by saying he has mathematically calculated *how near infinity can be for practical utility* — and it turns out to be at the other end of most of our basements if we stick to common mirrors! Here are the things that must be specified.

1. The desired mirror diameter.
2. The aperture to focal length ratio; *i.e.,* the $f/$number wanted.
3. The desired surface tolerance in fractions of a wavelength.

To go further we need to study the testing "setup," as shown much foreshortened in Fig. 8.5. The mirror at the left picks up the light from the source (S), an artificial star (pinhole) at the other end of the basement, and reflects it back to a focus at the knife edge (K) behind which a right-angle prism deflects it to the eye. Actually the mirror curve, after being figured to give a null test, conforms to a section of an ellipse, with the knife edge at one focus and the light source at the other, as the dotted elliptical curve indicates. Now if one were to move these two focal points farther apart, the mirror shape "approaches" closer to that of the paraboloid, and becomes one when our test source reaches infinity. All we need to know is how far we must move it to meet a desired fractional wavelength tolerance between our ellipse and a true paraboloid. The formula for calculating this as evolved by Spoelhof accompanies the drawing. To use this, one selects a distance for the light source (as your basement length) and calculates (for green light) the maximum separation or wavelength error between the ellipse this null test measures and a true parabola, where: our "error" or difference is in fractional, or whole, wavelengths; M equals Magnification, or light source distance divided by image distance; D equals the mirror diameter in inches; and the $f/$number equals the mirror focal length divided by mirror diameter.

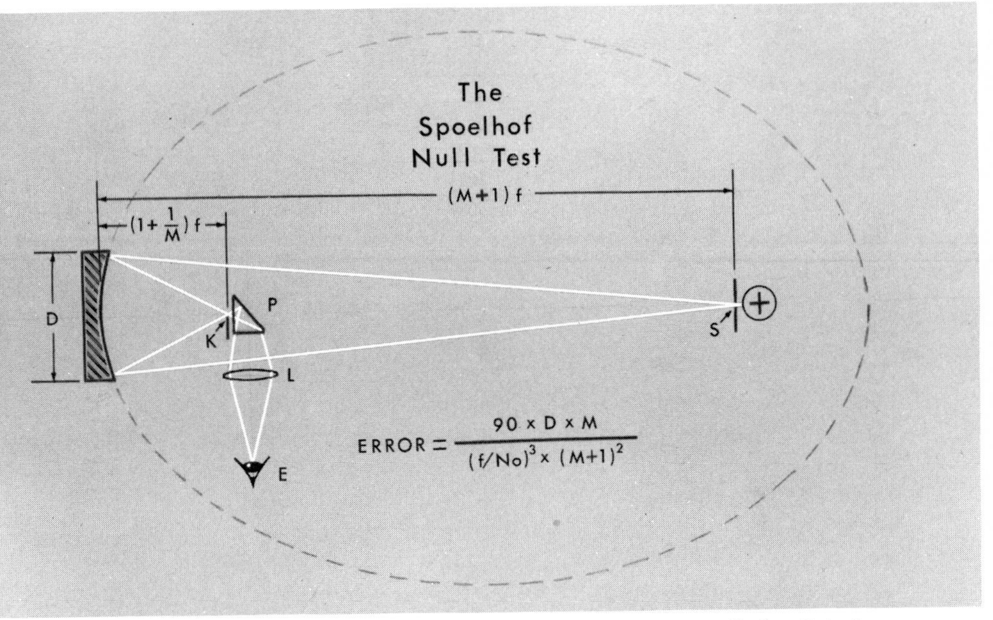

8.5 – Amateurs with a 40 ft. or longer work space can use the Spoelhof null test as described to construct highly satisfactory telescope mirrors of the most commonly used 6-in. *f*/8 type more easily. See text. Drawing by the author.

For practical purposes and reasonable source distances let the image distance equal the selected mirror's focal length, which in turn is half its radius of curvature, a value that can be found with a pen light and a piece of ground glass. The distance from the mirror to the source is (M + 1) x focal length. Now let's calculate one example and *you can then calculate any system you wish.*

For the most common 6-in. *f*/8, 4 ft. fl. mirror, let's assume we have a 40 ft. basement. Mirror to source separation = (M + 1) x 4 = 40, from which M = 9. Our calculation follows:

$$\frac{90 \times 6 \times 9}{8^3 \times 10^2} = \frac{4860}{512 \times 100} = \frac{4860}{51200} = .095 \text{ — or less than } 1/10\text{th wave!}$$

Here are a few *distances in feet* to place the test source to obtain the accuracies listed. Naturally, when you have *more* space, use it.

Mirror Data	Accuracy Desired		
	1/8th λ (.125)	1/10 λ (.1)	1/15 λ (.066)
4″–32″ fl. (*f*/8)	12′	16′	27′
6″–48″ fl. (*f*/8)	28′	40′	64′
8″–60″ fl. (*f*/7.5)	65′	80′	120′

These three tolerances might be considered as describing good, very good, and excellent mirrors. Where a *uniform* gray "cut off" occurs by this null test, the mirror is most likely to be of a known high degree of precision and will perform much better than one "thought" to be a paraboloid from first Foucault tests.

To complete and use the test setup shown in Fig. 8.5, a small piece of razor blade on a vertical wire stalk serves as a knife edge. Simple lateral and longitudinal control should be provided. A small prism (P) (enlarged in drawing for clarity) of about ¼″, as Edmund's 30,527, serves to deflect the light to a condenser lens (L), which in turn converges it to the eye, where the observer sees the shadows from the mirror projected onto this lens and thus does not have his head in the road of incoming light. A surplus 7 x 50 binocular lens or reading lens should serve, as long as it has a focal ratio half or less that of the mirror, say $f/3.5$ for an $f/8$ mirror, as provided by the binocular objective noted. The pinhole can be a fine fat needle hole of .01″ to .03″ in foil in front of a frosted bulb of suitable wattage, not the frustratingly small holes called for in Foucault tests.

Do not use a slit unless an accessory telescope is used to view the mirror. The small economical 50X pocket microscope with measuring reticle to a thousandth, as Edmund's No. 30,225, is invaluable for checking pinholes, etc. It should be noted that in an emergency you can effectively double the shop length with a *precision* optical flat somewhat over half your mirror diameter used at the basement end to "fold" the light path, placing the test source back near the test setup. Edmund's No. 60,450 — 4″ of ⅛ wave at $25 — might serve well, particularly if an optical shop friend would check it, since it's *got to be good.* The smaller size permits this emergency "basement stretcher" at a fraction of the cost of a full-sized collimator mirror.

MISCELLANEOUS HELPS

Adjusting the Polar Axis. Over and above what you read in directions and books, the following may help. For portable mountings attach to the head a medium-sized circular bubble level as made in useful types and sizes by Geier and Bluhm of 594 River Street, Troy, New York. In the type of mount where the head is rotatable in azimuth on the tripod or pedestal and the altitude has been previously adjusted, one can level the mount, and then with the tube set for 90° N, rotate the mount in azimuth until the pole star is picked up — quite a speedy operation. In a large scope with narrow field one may have to adjust slightly + or – from 90° according to location of pole star. A level instrument base is a good start for any type of instrument anyway.

As noted in A.T.M. Book Three, page 335, a small finder telescope, having a two-degree circle on the reticle (more exactly, twice the *present* Polaris from pole distance), which can be dropped down a hollow polar axis, makes a sublimely easy way to get your scope into adjustment for setting circle use. It can also be adjustably fastened to the side of a mount.

Filters can improve your observing. It's best to obtain a light to medium density series containing blue, green, yellow-green, yellow, orange and red and try these out on everything you observe. Kodak's No. 3N5 is fine for lunar work with either refractors or reflectors, being a combination of a 25% transmitting neutral filter and their No. 3 yellow. Many recommendations may be found, among which are the following readily available standard Wratten (Kodak) filters: Wratten red No. 25 (A), yellow No. 12 (minus blue), green No. 58 (B2) and blue No. 47 (minus red) are quite standard, but some lighter colors are useful, as lighter yellow No. 8 (K-2), light green No. 11 (X-1), and light blue No. 80. Why not buy cheap gelatin 2-in. squares and cut them up as needed, buying expensive glass mounted ones later for those most used. Excellent unwanted glass filters are often available by fine makers (Leitz, Zeiss, etc.) in "grab boxes" in large camera stores, at ridiculously low prices. Vernonscope and Optica b/c have a series of glass filters for astronomical use and a booklet on these.

Finders and cross hairs come in many forms. Usually two rings each having three locking screws at 120° bring the optical axis parallel to that of the telescope. It's more convenient if the pair at the top (away from the tube) is replaced with a strong pressure plunger screw-in unit, each constantly holding the tube pressed against the other two during and after adjustment. Finders can be either too low or high in power. The most useful range is from 4X to about 8X, but larger instruments may use higher powers, or better still, a low power 4–7X wide angle *plus* a 20–25X high-power finder. Fig. 8.6 shows a refractor fitted with both a conventional 6X finder and a fine 3-in. "Richfield" 12X wide-angle scope — wonderful to show first observers a fine wide field view of the area around a selected object. I like one with a right-angle eyepiece for reflectors, as shown in the section on accessories, Fig. 5.7. Finders too need dew caps.

Cross hairs of finders are usually illuminated by holding a flashlight "just right" near the front lens. If the cross hair reticle is glass, a ⅛″ hole can be drilled in line with the edge where a shielded flashlight nicely illuminates the engraved lines. The hole should be slightly back of the cross wire or thread type. The small grain-of-wheat bulbs with the rheostat on battery case power supply, as available from American Science Center and others, make a good reticle illuminator for either finders or guide scopes. Best is the new fiber-optic flashlight, no. 60648 from Edmund.

Greases and oils are needed, but what kind? For heavy pressure duty, as in slow turning worm gears, a non-binding grease should be used. The

8.6 — **A Richfield type refractor of about 3-in. aperture and 10 to 12 power, as shown here, makes a fine supplementary high-power finder. It presents splendid views of dim objects and introduces the beginner stepwise to objects seen at higher power in the main instrument. Richfield scope and photo by the author.**

best I've found is "Anti-Scoring Center Point Grease" from Chicago Mfg. & Distrib. Co., Chicago 9, Illinois. Do not use a light oil alone for cold weather but "cut" a grease with a light oil like "3-in-1" or even all weather auto oil. For large heavy bearings some mix in micro-graphite (from your bicycle shop) until good and black — it doesn't take much. Soap or "Door Ease" isn't bad for sliding aluminum and wood parts.

Bearings are probably proper if they suit your needs. Ball and taper roller bearings should at least be properly pre-loaded, assuming they are designed for this, to avoid play. They certainly add to ease of clock driving. Many prefer carefully made solid bearings, and tapered shafts on some instruments permit nice adjustment. Large solid bearings are not to be "sneezed at" for true rigidity, and those with nice wide pressure plates make for true sturdiness. Shafts could well be hollow to cut weight and heat storage. More shafts could be chromium, nickel or even copper plated to avoid rust. I have not used enough of the teflon and nylon materials, or these impregnated with graphite, for bearing surfaces to say more than that I'm sure they are worth considering in proper places. The thin teflon tape or sheet, gummed on one side like Scotch tape, certainly has fine bearing possibilities for either shafts or slides.

Metals in telescopes deserve one main comment: *don't* use ordinary steel if you can possibly avoid doing so. This applies to screws, knobs, gears or *anything*. I go far out of the way to get stainless, bronze, brass, tempered aluminum, fiberglass, bakelite, etc. Rust is a real nuisance with iron and essentially unavoidable, regardless of attempts at telescope care. At least where ball bearings are used heavy walled aluminum tube could well be considered for axes.

Heat Problems. Dew caps keep off dew, but they also keep body heat from boiling in front of a reflector tube in certain viewing positions. With open skeletal tubes a heat shield near the observer is desirable. Massive objects that store heat should be avoided around or near telescopes. Most people are aware of chimneys, but others are not so obvious. Once one of a matched pair of superb 6-in. reflectors performed poorly while its companion 10 feet away was resolving the limit. After much fussing and fiddling, it turned out that heat from my car engine 20 feet away was rising in the still air to ruin seeing in the one instrument pointed directly over and many feet above it. Trees and blacktop parking areas can do the same.

Be Comfortable. I'm lazy enough to want to be comfortable, but I do console myself with the known fact that keenness of observing for any protracted period is much enhanced if one is comfortable, as when looking slightly downwards instead of up at a neck-straining angle. Remote controls are really wonderful for even medium-sized instruments. Also, some company should make and sell a simple folding adjustable observing chair of wood or aluminum; many have been described over the years. All that would remain would be for you to sneak your favorite cushion out of the house. A binocular eyepiece (see Chap. 9) is wonderfully relaxing for observing the moon or planets; however, it's an impossible accessory for group observing, with its individual eye focus and eye width adjustments. Use sidereally driven slip ring circles and set only once in an evening. Why be cold with fine electrically heated suits available? We lazier members of the observing tribe never had it so good as right now. (See Fig. 3.18-left.)

Useful Powers. Concerning proper magnifications, resolving power, etc., many books supply needless tables and charts. A few comments should cover all. For practical purposes Dawes' Limit may be used where resolving power in seconds of angular separation is considered to be 4.56 divided by the working aperture of your telescope in inches. Accordingly a 2.3-in. (approx. 60 mm.) telescope will resolve two seconds of angle, a 3-in. one and a half seconds, a 4.5-in. one second, a 9-in. a half second, and so on. Some feel that small instruments in the 3- and 4-in. size range may exceed this, and most admit that atmospheric conditions or imperfect optics prevent large telescopes from ever reaching this, except on rare occasions. A certain minimum power, *about* 50 times per inch of aperture, must be used to capitalize on an instrument's resolving power. Somewhat more makes it easier for the eye to do so; however, powers much above *about* 50 per inch of aperture seldom give an added advantage, only making the image larger but displeasingly fuzzier and dimmer. I tend to use magnifications between 30 and 40 per inch of aperture at most, to enjoy relatively sharp images. What power for what object? Instead of consulting tables, start with a lower-power eyepiece and move up in power until the object appears best for *you* under the existing seeing conditions, the optical capability of your equipment, and the individual characteristics of your own eyes. This method is by far the quickest and best. Your instrument's power is always equal to the focal length of your mirror or objective divided by the focal length of your eyepiece.

For wide fields and dim object study we turn to the lower powers. Nothing is to be gained by less magnification than 3.5 per inch of aperture, since the exit beam from the eyepiece has now become as large as the eyes' maximum iris opening of about one-third inch, or 7 millimeters. To reduce power more is the same as cutting the aperture of the instrument. Often not recognized is the fact that as one becomes older the iris cannot open this wide, often making 4 or 5 times magnification per inch aperture more practical. Also, where there is even moderate sky light, as encountered anywhere near larger cities or broad suburbs, a smaller exit pupil of 5mm as obtained from a magnification of 5 per inch actually provides more pleasing views with darker sky backgrounds. A 30 to 40 mm. fl. eyepiece will provide some fine views with your standard 6-in. $f/8$ reflector. Edmund's 32 mm. No. 5160 with adapter No. 30,171 is a good one. Some manufacturers supply 40 mm. fl. eyepieces. Large aperture ratios, as $f/4$ mirrors or $f/5$ lenses, permit use of more normal focal length eyepieces for dim-object and wide-field viewing.

MISCELLANEOUS USEFUL ITEMS

Handy finders for medium and small-sized scopes are the rifle type, inverted V at front and notch at rear of the tube. For these a glow tape is ideal, since it glows for quite a period following exposure to a flashlight. One source is "Spot-Lite" tape from Lab. Supplies Co., P.O. Box 332, E. Station, Yonkers, N. Y.

Epoxies can be a scope maker's best friend. The long setting period is a handicap, but there is a *9-minute* fast cure, No. 309, from dealers for Hysol Corp., Olean, N. Y., who also supply many other types.

Epoxy finishes should also be considered. One fine example is the marine line "Poly Aqua" epoxies from dealers for Valspar Corp., Rockland, Ill.

Internal reflections in tubes and eyepiece holders can dull image contrast and lighten dark skies. The black flock paper from Edmund is fine for eyepiece tubes and especially for critical surfaces behind a Barlow lens. It can give trouble by "shedding" in tubes — so vacuum it after installing. An unusually effective flat black paint is the 3M Velvet Coating No. 9564 optical black. In refractors appropriate diaphragming should be used.

E. Ken Owen's "Blue Heaven" observatory captured through careful lighting. Note "tip-off" canopy, raised deck and elegance of instrument.

The 5-in. Celestron (left) is the best all-around size and weight (12 pounds). In my opinion, it is a best buy for an amateur. The 8-in. Celestron (right) has quite an aperture advantage, although still transportable (25 pounds).

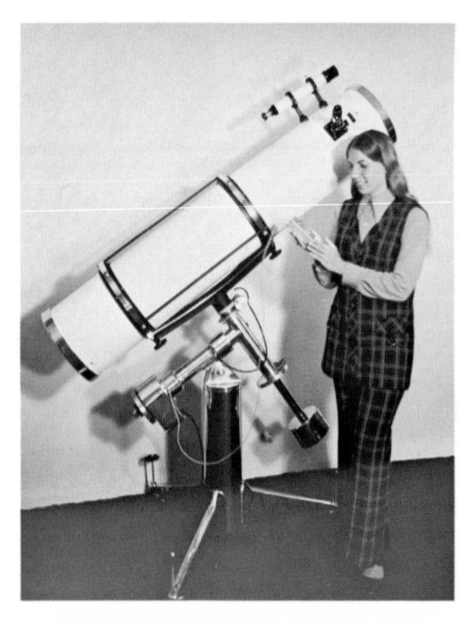

This Cave 10-in. *f*/6 Custom Super Delux is representative of fine commercial instruments. It has full remote controls with sidereal "slip-ring" right-ascension circle. Courtesy Cave Optical Co.

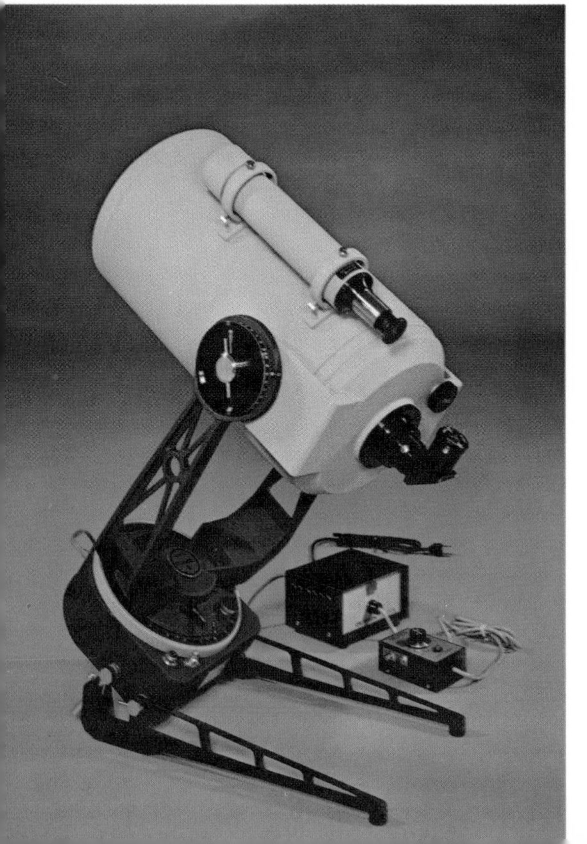

(Above) I designed, and Cave Optical Co. fabricated, this 12½-in. *f*/7 fused quartz mirror astrographic instrument. An angle, or "sheep's leg," pier permits clearance for large astro-cameras, replacing heavy counterweights on declination shaft. It is being placed into service by Mr. Nile Root, an assistant professor of Arts and Photography. Photo by the author. (Left) The more recent 8-in. Dynascope by Criterion—a fine-looking instrument by a good company. Similar "folded" optical systems are musts for photographing from a cramped spacecraft, as in the 1975 Apollo-Soyuz mission.

The planets' maximum apparent sizes, when nearest to earth, have been reproduced here at the same magnification. The planets' apparent diameters are measured in seconds of arc. Mars varies tremendously in apparent size as it approaches and recedes, as do Mercury and Venus. Saturn as it circles the sun, in approximately 30 years, presents the illustrated phase every 15 years. Now the rings are wide open. They will be edgewise in 1980 and will effectively disappear for a short period. Then the rings will slowly open again in reverse phase for about 15 years. Artwork by Victor Costanzo. Courtesy *Astronomy.*

(Above) An outside view of my Outer Space Observatory II reveals the upper side of the open 4 x 6ft., all-aluminum and fiberglass sliding, south shutter. The north side is essentially identical, serving as a windbreak for over 80% of my observing. Shutters open or close in approximately 10 to 20 seconds! (Below) Inside, wide-angle view with both shutters open. My favorite telescope, shown here, is a 6-in. *f*/10 refractor I assembled from a Jaegers objective, an old Cleveland mount, a sidereal Mark III drive (similar to a Byers) plus many modifications. Accepts diagonal binocular eyepiece unit (Zeiss) and even a 2-in. aperture 25X wide-angle eyepiece. The 5-in. Celestron works beautifully, as does the 80mm. (12X, 20X and 40X) Zeiss turret binocular.

9

BINOCULARS IN ASTRONOMY

BINOCULAR VIEWING

It has been said there is a proper instrument for every observing use, including each celestial object or sky area. Such an extreme stand is possibly a bit ridiculous, yet it's easy to present examples of specialized optics, as the 20X spotting scope for target or nature work, a superb refractor for lunar or planetary work, or a fine large reflector for *really* seeing the Hercules Cluster as a condensed ball of stars. When we come to viewing the Milky Way or certain wide-field star groupings on a pitch dark night, nothing beats a *good* binocular, one ranging from a 7 x 35 to a 10 x 80 in size. Higher-power binocular viewing of objects far out in space adds a new thrill to skygazing. It is known that there is a real physiological satisfaction in the relaxed viewing of objects with both eyes, even when objects are so distant that the stereopsis value of binocular vision has been reduced to nil. This desirable effect has long been recognized, Figure 9.1 showing an early high-power twin-tube French binocular telescope.

One way to enjoy searching through the Milky Way and looking at open star clusters on a dark night is to sit back in a low lawn chair with a high-quality 7 x 50 night glass — a most appropriate power aperture ratio for this use. If the night sky is black enough and you brace your elbows a bit, the views are simply amazing compared to naked-eye observation. Figure 9.2 shows the common binocular sizes I generally recommend. (The book *Binoculars and All-Purpose Telescopes,* H. Paul, Amphoto, covers in detail all aspects of choosing and using ordinary binoculars of all types — Ed.). Such an ordinary pair is a "first" consideration. The ten-power glasses are best tripod mounted — a camera tripod works well.

Some will enjoy their 7X to 10X glasses so much they will want to graduate to a pair of giant binoculars for a closer look at glowing nebulae and open star clusters. There is actually a most suitable type of glass, as

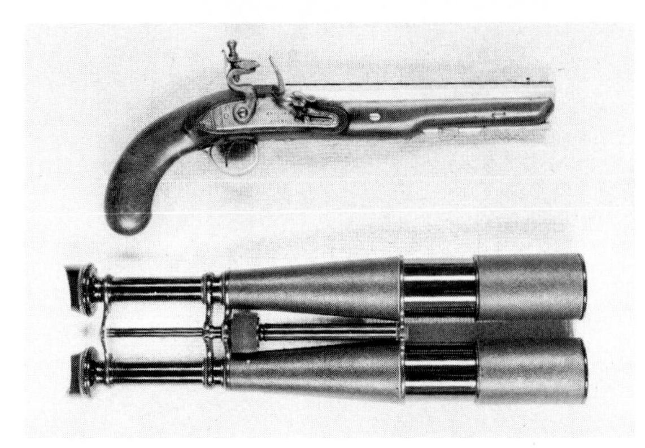

9.1 – **The long barrelled flintlock helps both date and indicate the size of this early French twin-tubed high-power binocular telescope. The advantage of binocular viewing in astronomy has long been recognized. Photo by the author.**

regards power, field of view and objective size, for each sky object. Most persons will be quite pleasantly surprised at how different the moon looks at 20 to 40 power when viewed with both eyes through a large binocular, or a binocular eyepiece on a telescope. One gains a real third-dimension impression, as though you could reach out and touch the moon, even though most of this effect is only a result of the natural relaxed manner in which one uses both eyes in stereoscopic viewing. Binocular eyepiece units may be readily attached to regular telescopes, as covered later.

LARGE BINOCULARS

A few of the more common giant binoculars ideally suitable for astronomical work and available as used or surplus glasses from military operations are presented.

Following World War II the German 10 x 80's — usually without the yoke mounting — were a drug on the market. Almost every camera or instrument store hoped to get rid of these large and heavy shelf-crowders. They must be collecting dust *somewhere*. These are marvelous astronomical instruments, the 45°-angle eyepieces eliminating most of the neck-kinking associated with binocular skygazing. They are the next logical step "up" from 7 x 50's. When tripod-mounted, these 10 x 80's, with their wide-field Erfle eyepieces and tremendous light-gathering power, are unbeatable for general viewing of the Milky Way, which stands out like a luminous ribbon of light studded with sparkling jewels. These are essentially "richest field" binoculars, showing more stars in the field of view than any other glass. A fine pair with original yoke mount is shown at the right of Fig. 9.3. Insert shows twin yokes, azimuth handle and elevation knob.

The large 20-power Japanese glasses also shown in this figure (left) are just right for looking at the large individual nebulae. The Pleiades is a long-to-be-remembered sparkling sight through even a commercial 15 x 60

9.2 – **Practical binoculars for general astronomical use.** (l. to r.) **Bushnell 7 x 35 mm., Leitz 10 x 50 mm., Bausch & Lomb 7 x 50 mm., and an old Busch 10 x 70 mm. marine glass. The first two are best near city lights, the latter two where it's really dark. See text. Photo by the author.**

binocular on a clear dark night, as is the dazzling double cluster of Perseus and many other exciting areas from one to three degrees in diameter. For full benefit of the giant glasses it *must be really dark* if you are to get full benefit from the true maximum light gathering glasses — those having a 7 mm. exit pupil produced by 3.5 times magnification per inch of aperture, as 7 x 50's, 10 x 80's, etc. Believe it or not, if there is much sky light, as near cities, you will not like the bright sky background and will actually enjoy more a glass having a 5 mm. exit pupil with 5 times magnification per inch — as quite ordinary 7 x 35's, 8 x 40, 10 x 50, 12 x 60, etc. Many have found this to be true, with the plus that these glasses often give sharper star images, are half the weight, and are much more available.

The many new 15 and 20 x 60's now manufactured with coated lenses are worth serious consideration. The optical coating can make these as efficient light transmitters as older uncoated glasses with 50 per cent larger objectives. Be sure that they are of suitable quality by star testing them to your own satisfaction.

The moon in all phases is always a most interesting sight at lower powers, even when viewed in a single-eyepiece telescope — particularly when it stands out sharply within about a 2–3° circle of black sky background. At the first sight of the moon through a 20X to 40X binocular many of my friends have remarked, "What a striking view!" The moon seems to be mysteriously suspended in space, and the mountains stand out as on a bas-relief map. I fitted a pair of high-power Erfle eyepieces to my 20 x 125's (Fig. 9.4) to take the power up to almost 40X. Such 40X or 50X binoculars are ideal for overall lunar observation, and particularly for your less serious viewers. Moon-gazers should always have at hand a series of cardboard aperture stops for the larger binoculars to cut out excess light

9.3 — **Many large ex-military glasses may be adapted for practical use in astronomy. Illustrated** (left) **a 20 x 125 mm. Japanese glass fitted to a sturdy elevating camera head.** (right) **A German 10 x 80mm wide-field binocular with original mounting, including finder and eyeshield. Note the double yoke and hand controls shown in the insert. Photos courtesy Paul W. Davis.**

and at the same time increase sharpness by reducing color error and lens aberrations. Resolving power will not be noticeably reduced if openings are not smaller than one inch for each 20 times magnification.

(*WARNING: Do not look at the sun without suitable sun filters and caution your friends accordingly.* These giant glasses are more dangerous than regular telescopes because of the smaller and therefore more concentrated solar images.)

There must be many of these old giant binoculars around "squirreled" away. To find one, simply use the same tactics you use in looking for stamps, antique guns, old autos or whatever else you may collect — ask *everyone.* The proper mounting of one of these for astronomical use is another story where a bit of guidance may help. The 10 x 80's can be mounted on a heavy-duty camera tripod with a tilt-top or pan head, such as a Quickset or Star D. This is by far the easiest way and reasonably satisfactory. Imported binoculars of similar aperture are also available on the current market.

The 10 x 80's of Fig. 9.3 have the original double-yoke mount (rare) used with these to support them at their center of gravity for easy swinging in all directions.

The larger glasses sometimes come with a yoke mount, as the 10 x 60

Japanese glass or the 15 x 80 size. Usually these have no mount and you are faced with using either a flat flange on each side, or a round trunnion, on which the glasses "hang" at the center of gravity in their original mounts, which were far too massive to be "liberated."

Figure 9.4 illustrates how one of these 20 x 125 glasses was mounted to swing at its center of gravity on a single cantilever arm support. The large flat "dog ears" on each side were simply sawed off, making a much trimmer instrument. This mounting is unsurpassed in its performance.

Figure 9.5 shows how binoculars with round trunnions on each side may be supported for easy sweeping of the entire sky. While this mounting is neat, light, and of aluminum, I've seen several made quite simply from heavy (full 1″) oak which, in my opinion, were steadier and performed better than my fancier job. Those who like shop work could build their own binoculars from surplus units, as shown in Fig. 9.6, at a fraction of their true worth. Such war optics are often excellent, having wide-field Erfle eyepieces at a convenient observing angle. How the one at the left was made is described in *Sky and Telescope,* pages 292-93, April 1957. It's actually notched into a bowling ball to make a universal mounting. The base extends below twice the distance shown. Binocular–scope units sometimes work well together, Fig. 9.7.

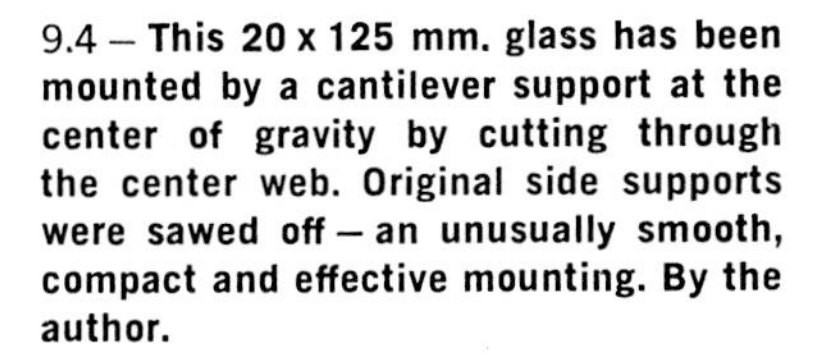

9.4 – **This 20 x 125 mm. glass has been mounted by a cantilever support at the center of gravity by cutting through the center web. Original side supports were sawed off – an unusually smooth, compact and effective mounting. By the author.**

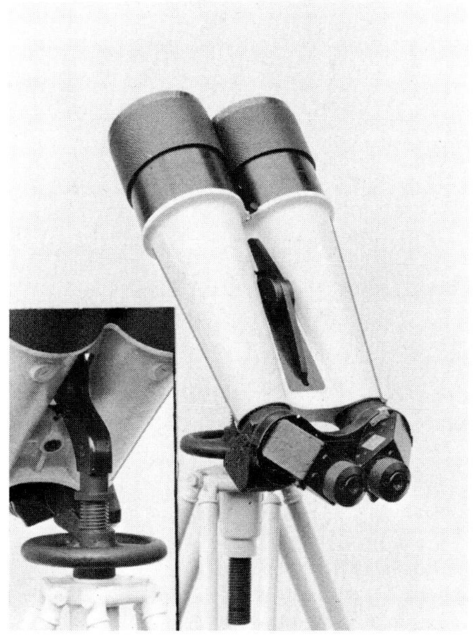

9.5 – **A giant 60-pound 15 x 125 mm. battleship night glass rests by its side arms on a home-built aluminum yoke. Easily maneuvered for countryside or sky observing. Heavy oak supports have served as well or better for others. By the author.**

If you are one who would really enjoy roaming leisurely around the blackness of a night sky for a closer look at the myriad of fascinating outer space objects, why strain and squint through a single low-power eyepiece when a good binocular is so much more pleasurable?

BINOCULAR EYEPIECES

Those who have a 3-inch or larger refractor of adequate light capacity will be amazed at the ease and pleasure of studying detail, particularly of the moon and planets, through a binocular eyepiece attachment. These have not come into common use, possibly because of manufacturing difficulties

9.6 – (left) **A fine pair of high-power binoculars from a mated pair of surplus angle telescopes with added long-focus objectives, as built and photographed by Paul Shaad.** (right) **A pair of large 12 x 80's assembled by the author from surplus optics and parts – with some machine work.**

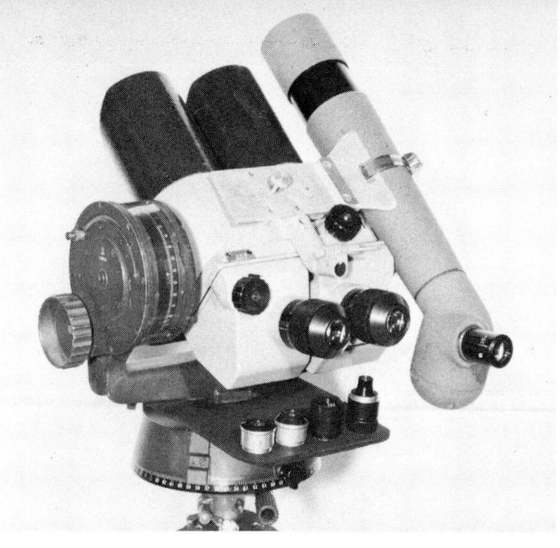

9.7 – **A binocular-scope teaching unit: the 10 x 80 for location and Richfield views, and the Bushnell angle spotting scope with Barlow-eyepiece providing powers from 15 to 120X for study. A practical unit, ideal for armchair astronomers. A B.&L. zoom 15-60x Balscope with 45° eyepiece, despite price, would do an outstanding job on such low power binoculars. By the author.**

and expense. In Fig. 9.8 is seen an old commercial unit (Carl Zeiss) I found and had "coated." Mechanically inclined amateurs can easily pick up a used microscope binocular eyepiece unit, the inclined ones being especially good. Some even use the microscope eyepieces with them. In the figure is shown such an ancient Bausch & Lomb unit from a pawn shop hastily adapted to a star diagonal for real ease of viewing. The eyepieces are regular focusing 22 mm. Kellner surplus from military 6 x 30's taped on.

A hint or two. Such units adapt best to refractors. Some may require a little tube shortening to make up for the "optical length" of the added unit. Reflectors present greater problems. The mirror may be moved up the

9.8 – (left) **A binocular eyepiece, as the old rejuvenated Zeiss shown here, adds much to lunar and planetary observation with 4-in. or larger instruments.** (right) **A hastily assembled binocular eyepiece made from a pawn shop supplied ancient B. & L. microscope unit with taped on 22 mm. surplus focusing binocular eyepieces! Works perfectly. By the author.**

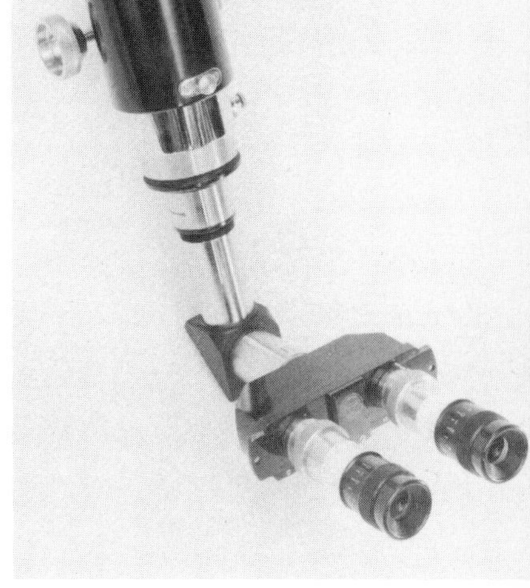

tube, or the tube shortened. A Barlow may be used at the tube wall to bring the focus out farther. This changes the *f* value to 15, 20, or more and is good for high-power work only, as for planets. Another method is to obtain a pair of small achromatic surplus *objective* lenses 25 to 30 mm. in diameter and 4 to 5-in. fl. These can be placed together (most curved surfaces facing each other) at the front of the binocular eyepiece unit 4″ to 5″ behind the telescope's focal plane in a separate holder to gain working distance (see Fig. 5.4g). Such a scheme also serves as an erecting system for terrestrial viewing!

9.9 — (left) **For superb wide-field night sky views use a 7 x 35 wide-angle glass, as this 11° W/A Rangemaster.** (below) **A 20-power binocular on a tripod permits both astronomical and terrestrial viewing. Several companies make 20 x 60, 70, or 80 mm. binoculars, but buy only on a trial basis. Photos courtesy D. P. Bushnell Co.**

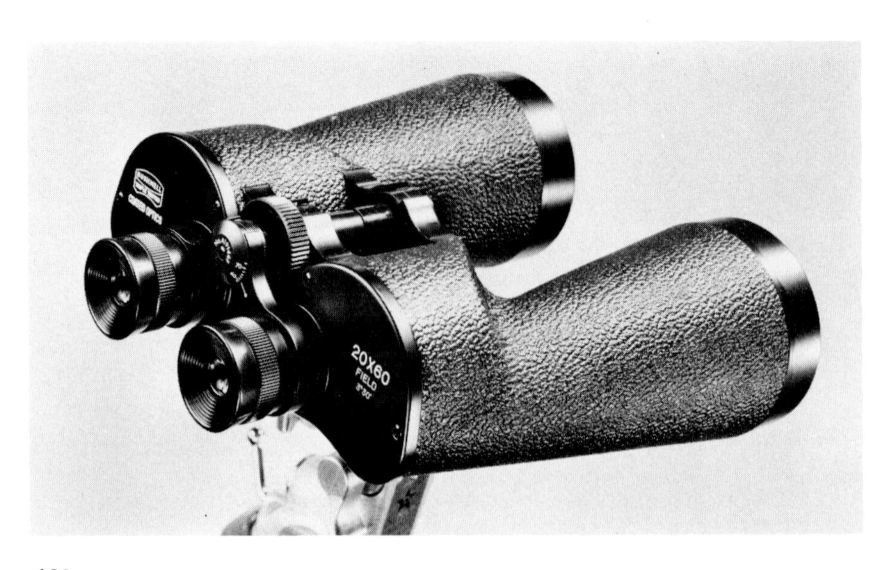

10

INTRODUCTION TO ASTROPHOTOGRAPHY

GENERAL INFORMATION

You can photograph anything you can see. In fact, the photographic emulsion can in most cases go far beyond the eye, having the advantage of being able to store or add up light's impact over a time period to yield strikingly detailed pictures of objects appearing only as a dim haze to the eye. A good example is the tremendous whirling galaxy Andromeda, as seen in the outstanding amateur astrophoto by Dr. C. P. Custer in Fig. 10.1 (pages 138 and 139). To the uninitiated observer his first telescopic view of this object is a great disappointment — usually eliciting a remark like, "*That* dim hazy patch is Andromeda?" You then hasten to explain what photographic film can do that your eye can't.

Any telescope or lens may be used with a film holder to become an astrographic camera; however, some types are better than others for each subject or area. Some fine examples of amateur astrophotography are presented throughout the chapter, to show what can be accomplished. First, let's look at the more common ways to use lenses in astrophotography, as illustrated in Fig. 10.2. At (A) the light from the telescope's main lens, or mirror, falls directly on the film as it would from a camera lens. This is usually called photography directly at the *prime focus;* the telescope is without an eyepiece and the camera without its lens. At (B) the camera with its lens is placed snugly up against the eyepiece of the telescope, with *each* instrument focused at infinity. The size of the image on the film is proportionally larger as the focal length of the camera lens is longer than that of the eyepiece. At (C) the eyepiece is used like a projector lens to throw the image from the main telescope or mirror onto the film in its holder, or camera box without lens. The final image size again is proportionally greater by the value of the lens-to-film distance divided by the lens-to-primary-image distance; *i.e.,* longer rear projection distances yield

10.1 – This splendid photo of the great spiral galaxy of stars known as Andromeda was taken one part at a time in three separate 2½-hour exposures by C. P. Custer, M.D., and reassembled as a photomontage. Procedure under photographing nebulae.

larger pictures. This is a favored way for high magnification work. One can also obtain moderately larger sized images at the prime focus and make this more accessible in reflectors by placing a Barlow lens in the cone of light from the lens or mirror, as shown at (D). Criterion makes a fine unit to attach 35mm cameras to *either* reflectors or refractors, as shown in Fig. 10.3 (left). When using a Barlow system you are essentially converting your optical system into a common telephoto type lens unit. How to calculate image size is described and illustrated in Fig. 5.3.

Also shown in the figure is a Bushnell adapter for their spotting scopes. It could, with a little machine work, be adapted to fit and clamp over the eyepiece onto the eyepiece holder, yet permit exchanging eyepieces — or make your own. This together with a suitable pair of *orthoscopic* eyepieces, as *about* 8 and 24 mm. for reflectors and 12 or 16 and 32 mm. for refractors,

would permit most usable equivalent focal lengths, *f* values and image sizes. Most telescope makers have some form of camera adapter. Such a unit is shown at the right of the figure, where a modified Nikon is coupled to a Questar 3½-in. lens–mirror scope. I've seen many remarkably sharp astrophotos of brighter objects made with this unit, a compact one that can double for nature photography. Spacek Instrument Co. sells a fine equatorial mounted and driven "Camera Platform" with guiding controls (and guide scope too if wish) — just attach your camera plus whatever lens you select.

What about lenses, telescopes and cameras? Only a few comments seem appropriate in this brief chapter, and I'll omit the usual pictures of lenses, mounts and cameras (covered in more detail in *Outer Space Photography for the Amateur,* Amphoto — Ed.) in order to show you a little of what amateur astrophotographers *can* do and tell how they did it, insofar as data is available. I'd like to assume that for even early serious attempts an electrically driven equatorial mount is at hand, with at least a simple guiding telescope having a cross hair reticle of some sort; even though star trails, meteor and aurora photographs, and appropriately short exposures

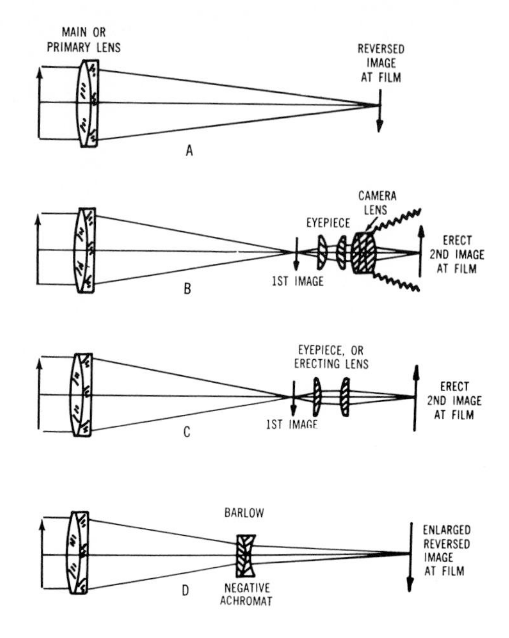

10.2 – Shown are the four most commonly used procedures in astrophotography. (A) **Photographing at prime focus,** (B) **Entire camera coupled to eyepiece,** (C) **by eyepiece projection directly on film method, and** (D) **the negative Barlow or telephoto type system. From** Outer Space Photography for the Amateur.

of the stars, the moon and the planets on ultra-fast films in stationary cameras can produce valuable photographs.

The single-lens reflex camera with interchangeable lenses (total lens, not just front element) is really the answer to the astrophotographer's dreams. It's by far the quickest and best way to get into astrophotography. With standard lenses of *f*/1.4, *f*/2.0, or even *f*/2.8 and fast color films many striking celestial photos can be made with the camera attached to the side of your scope, and occasionally when it is tripod mounted. Using this type of camera, you are always able to locate objects in the field quickly, and what is more important, determine that they are in sharp focus. With or without its lens, an astronomical telescope may be used with one or more of the methods just described to obtain the desired image size. Although somewhat restricted in this application, reflex cameras with fixed lenses can be used in method B, discussed above.

What cameras? Since the supply is virtually endless, may I recommend a few to consider, somewhat in order of my preferences in low and high cost groups. Among the economically priced is the Miranda, Praktica FX, Edixamat, Pentax and Praktina. The large opening of the Miranda reduces vignetting, or loss of edge-of-field illumination, from long focus systems and is my favorite for the money invested. Their *f*/2.0 50mm and *f*/2.8 35 or 28mm lenses are fine (see Fig. 10.3). Most other lenses can be adapted to it. The Praktica FX can be a lowest-priced good buy (particu-

10.3 – (left) **Criterion and Bushnell units to couple single-lens reflex cameras (without camera lens) to telescopes. The Miranda shown is one of the cameras recommended. The Honeywell Pentax is also excellent, since it has a widely used screw lens mount.** (right) **The superb Questar telescope is shown here coupled to camera ready for astronomical or long-range terrestrial photography. Photo courtesy Questar Corp.**

larly a used or earlier model)with an exceptionally standard lens mount to which "used" bargains can be adapted. Among the higher-priced is the Leicaflex (lenses are superb) and the established Nikon F or Nikkormat the Exakta and the Canon Reflex. All of these cameras are amply illustrated in current camera magazines. Where feasible, a clear glass-crossline unit should by all means replace ordinary ground glass or other screen — at least in its central spot. All focal plane shutters produce some vibration, and return-mirror cameras the most (unless the mirror locks up). Whenever possible make exposures of a second or more using the old-fashioned card-in-front-of-lens method. For large scopes some cut a slot in the tube in front of the camera to do this, others fit a relatively vibrationless large old Compur shutter in front of the focal plane unit for exposures of one second or faster — quite a fine solution when combined with the card method for exposures over one second. Many marvelous lenses are available from these makers, or from others to be readily adapted, and outmoded used ones without automatic diaphragm, as *f*/1.5 Biotars, are wonderful buys. Sharper lenses are available, but at 5 to 10 times the price! Your camera store will supply literature on any or all. The 35mm camera permits the widest possible selection of black and white and color films, and even some of Eastman's special films for dim light photography. With the longer focal lengths for wider fields it's often desirable to turn to larger film sizes. The 2¼ x 2¼ is probably as large as is practical because of the need to maintain film flatness and accordingly sharp focus. Cameras like the Bronica, the Hasselblad, Praktisix, or Reflex 66 can be adapted, as well as old camera backs with high-quality cut film holders. Large cut films without a vacuum back are not recommended. Many who wish larger negative areas turn to the more precise photographic plates and make their own plate holders and ways of checking the focal plane. Such can be quite simple. Then you can use Eastman's superior emulsions, widely available on plates, like blue sensitive type 103aO and high red sensitivity type 103aE or newer type 081-01, etc., as listed in the booklet "Kodak Photographic Films & Plates for Scientific & Technical Use." Such special plates are highly recommended for aperture ratios of *f*/4 and slower in larger instruments at prime focus for photography of the fainter diffuse sky objects. Store all film in sealed containers in your deep-freeze, or at least in a refrigerator.

Ordinary photographic type lenses, besides current commercial ones, are gathered by most astrophotographers from war surplus. Widely used is the Eastman *f*/2.5 Aero Ektar in 7- and 12-in. focal lengths (see lower Fig. 1.12 for a photo by one). Use wide open for dim diffuse objects; at f/4 for meteors, etc. and for *moderate* definition and at *f*/5.6 for really sharp work. The Ross Express *f*/4 lenses of 5-in. and 8¼-in. are good buys. The Bausch and Lomb Metrogon 6-in. wide-angle survey *f*/6.3 (*used at f/8*) is one of a very few lenses that can cover *about* 9 x 9 inch (5 x 7 better) to produce fine star photos. (See meteor photo Fig. 1.16.) In longer focal lengths the 24-in. *f*/6.0 Bausch and Lomb Aero-Tessar and Eastman Aero Ektar are very popular. Shield's photo, Fig. 1.13, was taken with one. Use long focus surplus lenses with a yellow filter — near film unless of highest optical quality glass. Always buy larger lenses subject to trial, since they can vary greatly. Broad experience teaches that, except for certain slow wide-angle lenses like Metrogon, an area of 4 x 5 inches is about all that can be adequately covered with reasonable definition by *any* of the lenses (or even telescopes) most amateurs are likely to use. It's a readily available size and about the limit of practicality for enlargers. Naturally there are no limits in any direction for the more experienced, but the above may save some disappointments. Sources of large lenses: war surplus dealers, Edmund, firms that advertise in *Sky and Telescope* magazine, used lens departments of large camera stores, or by advertising.

Most refractors with a yellow filter near the focus can do fine jobs on the brighter objects, such as the moon, sun, planets, brighter stars, etc. For dimmer objects a 6-in. or larger reflector of *f*/8 or faster is highly desirable. Photographing directly at the prime focus (as shown at Fig. 10.2A) is best for nebulae, galaxies and fainter objects, while a projection system as shown at (C) is most often used for planets and lunar or solar "closeups." Here again an aperture of 6-in. or larger is preferable for serious effort. The

Criterion combined Barlow unit and adapter to attach your 35mm directly on a standard reflector, as illustrated earlier, is a nice addition. If you have a purchased instrument, always write the maker for advice and guidance, or build your own astro camera — it's fun. For more detail please see *Outer Space Photography for the Amateur.*

I feel it's a real help to study other workers' accomplishments and have therefore asked several who take fine astronomical photographs to present their work. Where available, basic photographic data will be supplied, but great variations in "seeing" conditions, or lack of one piece of information, makes such data best used primarily as a reference guide. I believe the illustrating of top-ranking work by other amateur astrophotographers will go a long way to show what actually can be done with perseverance and meticulous care to all details, as well as challenge you to do better. As proud as I've been of several fine photos I've worked quite hard to get, each year I see a similar photo — *only better.* If you work with newer films, developers and techniques, you too can surpass those presented.

A word on *films* or emulsions, the next most important thing after lenses. First, using the right film *is* important but stating the right film for each job and condition is almost impossible, as you will later see. Figure 10.4 strikingly illustrates that a *high contrast* film is a "must" for full moon photography. Note the tremendous difference in results. Kodak 35mm High Contrast Copy film or Adox Dokupan would also be fine for this. But on partial phases of the moon at the shadow line, where contrast is high already, Plus-X would be quite satisfactory. As an example in the

10.4 – **These two photos strikingly illustrate the use of right and wrong types of film for photographing the full moon. A high contrast film must be employed, as used at the left. Courtesy John T. Hopf.**

area of planetary photography, particularly with smaller-aperture instruments, the small image even with ample projection makes one want to choose fine-grain (slow) film. However, unless seeing conditions are quite good, the atmospheric image motion over the exposure time required could really mess up your image. A much shorter exposure on a somewhat grainier but faster film could yield a better picture. I'd recommend a medium or fast film, as Kodak Plus-X or Royal Pan and projection by eyepiece to provide equivalent *f* values of from 64 to 256. Calculate the equivalent *f*/number by dividing the projection distance by the eyepiece focal length and multiplying this value by your scope's *f* value. For an *f*/8 reflector with an 8 mm. eyepiece and the film 7″ (175 mm.) to its rear, we have the equivalent of an *f*/175 system with an image 22 times larger than at prime focus. Dennis Milon has made some superb planetary photos with Royal Pan at his recommended *f*/value of 170 (12-in. *f*/8 reflector). Yet at prime focus, as for full image of a gibbous moon, a slower finer-grain film, as Adox Dokupan, is needed to best utilize your instrument's resolving power. More about this later.

Longer exposures beyond a minute or so *do not* yield proportionally denser images because of a phenomenon called "reciprocity law failure." Films vary, and Eastman Kodak can provide you data as needed. In their booklet on special films and plates previously noted, one will find materials especially designed for long exposures in dim light. Observe storage and shipping requirements. Color film types are constantly changing. On long exposures marked "failure" occurs, and to make matters worse it differs for different colors, objects no longer appearing in their natural color. Try some and pick the one *you* like best. George Keene, myself and others feel High Speed Ektachrome is a good starter film. Try Kodachrome-64 and other fast color films. Get Kodak's list of color correcting filters to end variable "failure" for different colors. With color, the faster the lens system the better the results, hence *f*/1.4 to *f*/2.0 is best with standard 35mm cameras, and Schmidt cameras should be considered by builders.

For broad general information a table from *Outer Space Photography for the Amateur* (Table 2) is reproduced as a starter.

PHOTOGRAPHIC ACCOMPLISHMENTS

Astronomical photos will be presented here to follow the areas as covered in Chapter One, along with such useful data as are available. A few comments are sometimes in order, over and above the basic data. At the end of *each area* fuller information will be supplied on some of the photos of Chapter One, but only where it contributes new information.

Solar Photography merits a few do's and don'ts. Never place the camera at the direct focus of any but the smallest of telescopes or well stopped

TABLE 2

OBJECT-INSTRUMENT REFERENCE TABLE

Object	Instrument	Lens	Focal Length	f/number	Comments
STAR TRAILS	Your camera.	As available.	Any — i.e. 2″—12″ but 5″ to 8″ good	Wide open but not faster than f/3.	An f/4.5 lens covering 40° best all around.
SATELLITES	35 mm up to 4 x 5 cameras.	f/4.5 or faster.	2″ to 8″.	Wide open.	40 to 60° angle of medium coverage okay.
MOON	Camera or scope + camera	Reflector or refractor or teleph.	48″ to 100″, but at least 20″.	f/8—f/15: by amplification to f/60	35 mm reflex camera + reflector or refractor ideal.
SUN	Camera + filter*; also + telescope.	As above. No filter at total eclipse.	As above. 3″ aperature + for detail	f/15 to abt. f/60 (totality — open)	As above — amplif. lens for desired magnification.
STARS (general)	Camera. (wide angle work)	1″ diameter min.—preferably larger	Any — but 5″ to 12″ best.	f/4 to f/8 as req. for definition.	Medium speed lens, or stop, for wide angle definition.
STARS (clusters)	Camera + scope (narrow angle)	2″ dia. minimum—preferably 4″ +	20″ or longer	f/6 to f/15.	Long focus for greater magnification.
NEBULAE	Camera or camera + telescope.	Scopes 2″ or more lenses 1″ or larger	5″ + for cameras. Short focus scopes	Camera f/2 to f/4.5 scopes f/5 to f/8.	Need greatest obtainable light grasp.
COMETS	Camera (+ scope if comet small)	Scopes 2″ or more lenses 1″ or larger	As for nebulae.	Usually wide open.	Good light grasp and angle to cover object.
AURORA	Camera.	1″ dia. or larger fast lenses.	2″ to 8″	Lens wide open. f/2 to f/4.5 best.	40—50° angle fast lens.
METEORS	Camera.	As available.	2″ to 8″	f/3 to f/4 best.	40° angle of good coverage as by Ross f/4 W. A. Express
PLANETARY	Scope + camera.	for detail 6″ + (minimum 3″)	Equiv. focus of 50″ to 1000″.	f/15 to f/100 equivalent	6″ or larger reflector + amplification lens best.

* Filter density of 4 or 5 to reduce light 10,000 to 100,000 times—no filter at totality—do not look at sun thru scope.
+ = "plus" or "or larger" as fits the case.

down lenses, or a burned-out focal plane shutter cloth, or even an injured eye can result. Reduce the incoming light by one or more of the following: a solar wedge, suitable neutral filters, an unsilvered mirror or diagonal, a projection method, or appropriate stopping down of the lens in some cases. Gelatin filters should be in front of lenses — not where it's hot. Gelatin neutral density filters are inexpensive for instruments of 4-in. or smaller aperture. Start by getting densities D4 and D5 (one ten thousandth and one hundred thousandths transmission respectively). Try Verichrome with D5 filter, 1/60th at *f*/11 as a starter, and take it from there. Larger instruments must use other means, as thoroughly covered in Chapter Five. Eastman's free pamphlet "Solar Eclipse Photography" contains useful information.

In Fig. 10.5 is an unusually beautiful view of a flaming rayed outer corona. Three seconds at *f*/13 with 3-in. refractor, Panatomic-X in D76. At totality the light is remarkably dimmed and fast wide-latitude films are often used to catch the very faint outer coronal streamers. Getting full tonal gradations from film to print is quite a challenge. Long-focus lenses can also yield unusual pictures of a sun squatting on the horizon behind interesting silhouetted foreground objects. This is an unusually fertile field for color. I've seen some color photos centered about the sun that rivaled any painter's work.

Figure 1.2, the partial eclipse sequence, was taken at 10-minute intervals, fixed mount, with a 3¼ x 4¼ Speed Graphic 170mm Kodak anastigmatic, 1/400 sec. at *f*/22, E. K. Contrast Process Pan, Wratten A plus 5% transmission neutral filters. The annular eclipse 1/25 sec. at *f*/11, 40-in. equivalent focus, Kodachrome (ASA 10). Diamond ring, at focus of 640 mm. Novoflex lens, ¼ sec., at *f*/22 on Plus-X film. Totality 1 sec. at *f*/4.5 with 210mm Zeiss Tessar on Adox Dokupan. Sunspots with a 6-in. refractor. Prominences were through a Hα monochromator and a special setup too complicated to describe. The outstanding single sunspot photo with one second granular detail was photographed through a Questar telescope with long eyepiece projection.

Lunar Photography intrigues many. It's an ideal starting point, yet one with real challenge when maximum resolution, detail and tone gradation are desired. Films, developers and papers are very important to those who are masters of lunar photography. The striking photo of Fig. 10.6 with the bright sharp lunar crescent touching the treetops and the cradled moon glowing duskily from "earth shine," shows that extreme magnification is not needed for esthetically pleasing photographs. Exposure 1½ sec. with 49-in. *f*/7 lens, equivalent of blue filter, medium speed film. In Fig. 10.7 are four of the finest "closeups" of the moon it's been my pleasure to see. Only 5 x 6″ portions of superb 8 x 10's could be presented at about two-

10.5 – During solar total eclipses the glowing corona streaming far out from the sun may be photographed, as shown in this fine photo by Roland Rustad, Jr., taken with a 3-in. refractor.

thirds original scale. Top left: 12½-in. Cave reflector, eyepiece projection to *f*/60, ½ sec. on Kodachrome-X and originally enlarged 12 times. Top right: same instrument, eyepiece projection to *f*/70, 1 sec. on Panatomic-X, FR X-22 developer. In Fig. 1.3, this area is pointed out on a last quarter photo. See Jan. 1965 issue of *Sky and Telescope* for full details. Lower left: 8-in. Astrola reflector, 8 mm. eyepiece projection to *f*/145, 3 sec. on Royal Pan, developed normally in DK60a. Lower right: 12½-in. reflector, 8 mm. eyepiece projection to *f*/170, 1 sec. on Royal Pan, developed in DK60a. Orthoscopic eyepieces always used. These last two outstanding photos taken before first quarter resolve craters 1.5 miles or less in diameter. Mr. Milon feels his present position in professional astronomical photography may well have come about from his accomplishments in lunar and planetary photography as an amateur.

To obtain best results in lunar and planetary work there should be a proper coupling of film, *f* value and exposure time — *and* excellent "seeing" conditions. Film resolving power in lines per millimeter is always based on high contrast charts; however, planetary and faint lunar surface gradations are quite soft in shadings. For this and other reasons it's best

10.6 – A 29-hour old-moon hangs as a silvery crescent above the treetops, its face dimly illuminated by sunlight reflected from the earth. Photo by Alan McClure with a 49-in. focus f/7 lens.

to use a film having two or three times the resolving power of the system (*f* value) used. Pope and Osypowski, judging from both theory and practice, recommend the following *f*/number–film combinations; *f*/20, Adox Dokupan or Kodak High Contrast Copy; *f*/40, Kodak Panatomic-X or Adox KB14; *f*/80, Plus-X or Tri-X; *f*/160, any film, since most high speed films, as Royal Pan, will then have ample resolving power at this *f* value (note this latter value ties in with Milon's findings). What this amounts to is that if you use a fast (low resolving) film at focus (as *f*/8) or any projection value faster (larger aperture ratio) than *f*/160 you are not getting the benefit of the full resolving power of your telescope or scope projection system for the smaller image size you've chosen. If you use a slow (high resolving) film on systems slower (smaller aperture ratio) than *f*/20, you are wasting the resolving power of the film and unnecessarily increasing the exposure time for atmospheric or image motion problems. Hence the balance between image size, aperture ratio and film character-

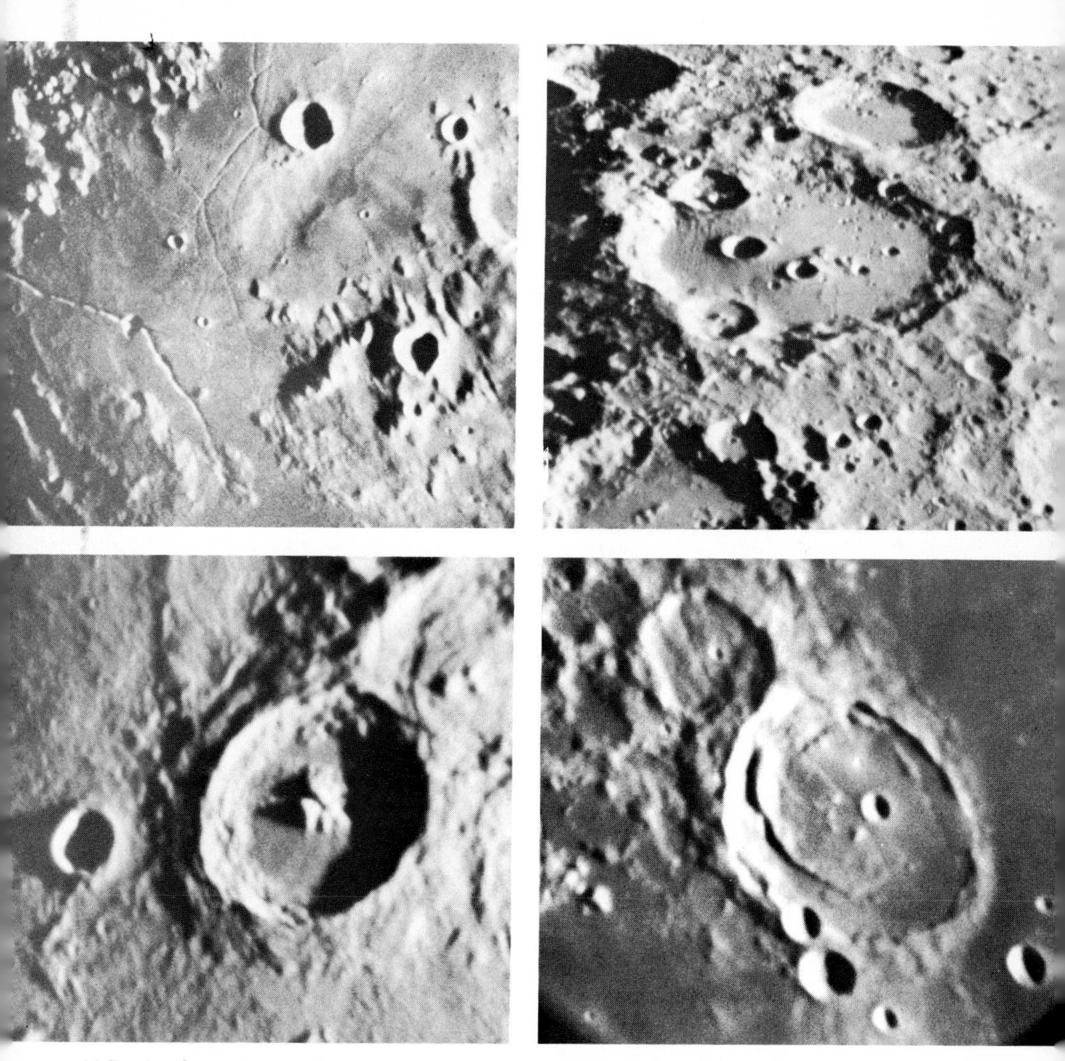

10.7 – **In the order presented, these superb closeups of the moon show the spidery Huygens clefts, one interrupted by a crater, by T. Osypowski; the Clavius area of craters within craters, by T. Pope; sixty-one-mile diameter Theophilus with its 7300 ft. central peak, by D. Milon; Posidonius and its peculiar ridging, photo by Paul Knauth, print by D. Milon. First two photos – 12½-in. reflector, last two – 8-in. reflector. See text.**

istics *coupled* with exposure time, seeing conditions and the character of the surface or object being photographed. For the beginner a mid-point, as characterized by Plus-X and Kodachrome-64, is a good place to start; then move in a direction to meet your needs, after mastering several things lightly glossed over here, such as good optics, perfect focus (using clear glass–cross hair method), accurate guiding on the object, excellent seeing and proper film processing and printing techniques. Outstanding astrophotographs don't "just happen." Guideline exposure times reported by Pope and Osypowski for a gibbous moon are as follows: Plus-X, 1/250 sec. at *f*/9 and ¼th sec. at *f*/90; Kodachrome-64, 1/60 sec. at *f*/9 and ½-1 sec. at *f*/45. Others can be estimated.

Lunar eclipses should be taken in color, and some eclipses, because of atmospheric conditions, are exceptionally beautiful in their deep rich colors, capturing all parts of the yellow–orange–red end of the spectrum. Since black and white copy misses the lunar eclipse story, I've omitted a series on Kodachrome-X of a colorful total lunar eclipse; however, exposure times and data for my best photos with small *undriven* equipment were as follows, as a rough guide. Bushnell spotting scope set at 30″ equivalent focus *f*/16 at about ⅓rd of totality — 1/15 sec.; at about ⅔rds — ⅛ sec.; small crescent of direct illumination left — 1 sec.; early stages totality 8-in. fl. lens at *f*/5.6 — 4 sec.; darkest phases 4-in. fl. at *f*/2.8 — 10 sec. The moon varies markedly in illumination near its darkest point from one eclipse to another. With driven *f*/8 reflectors and color films, as Kodachrome-64, I'd recommend starting with one minute exposure and doubling this up to 16 minutes or so. You should have one photo with fine red coloration.

In Figs. 1.3 and 1.4 we have two other outstanding examples of lunar photography. It's unfortunate that these masterpieces of shadow gradation and superb detail just can't be reproduced to do full justice to the originals. The first was with the Cave 12½-in. reflector at *f*/8 (prime focus), ⅛th sec. on Adox Dokupan — 15 min. in FR X22: 1 to 9, and the second was identical in conditions to that of the Huygens clefts just shown.

In *planetary photography* we can apply almost directly what we have already learned from high resolution lunar photography by the projection method; little new needs to be added. Except for lunar eclipses, planetary photography has more to offer the color photographer. Here Kodachrome-64 is again recommended. First check your "seeing" conditions carefully, particularly as regards image motion, before starting to avoid wasted time. Except for exploratory photos, the image should be kept centered for utilizing best optical performance. Here is a good place to experiment with color filters with black and white and correction filters with newer color films. Osypowski and Pope of the Milwaukee Club offer the following ex-

10.8 – Five excellent planetary photographs. As presented: the Crescent phase of Mercury has just started in this photo by H. Dall; the polar cap and delicate markings on Mars by A. Dounce; Saturn print from combining two negatives by T. Pope; transit of satellite Ganymede's shadow across Jupiter by D. Milon; unusual belt fading near the Red Spot, by D. Milon and D. Burkes. Photo projection methods with 8- to 16-in. reflectors.

posure guides: For Plus-X and Jupiter — ½ sec. at *f*/90, and for Saturn 2 sec. at *f*/90. For Kodachrome-64 and Jupiter — 4 sec. at *f*/90, and for Saturn 8 sec. at *f*/54. For finer-grain Dokupan and shorter projection distances, with Jupiter 2 sec. at *f*/32, and for Saturn 8 sec. at *f*/32. From this one can calculate for other *f* values and estimate for other films. All photos presented are by the projection method — usually with 8 mm. orthoscopic eyepieces.

In Fig. 10.8 are some outstanding planetary photos. The first of Mercury by my esteemed friend Horace Dall of Luton, England, is quite unusual. The planet was only 7.9 seconds in angular diameter and the phase concavity less than one second. It was photographed with a 15-in. Cassegrainian by projection to an equivalent focal length of 13,000 inches, or at a focal ratio of about *f*/900. Mars with its polar cap and shadowings was photographed by Dr. Dounce with only an 8-in. $f/8$ reflector; however, this instrument was a prizewinner for optical perfection, again indicating the importance of quality optics. Saturn, as photographed by Tom Pope, was printed from two superimposed negatives to reduce grain. Four second exposure at $f/90$ on Panatomic-X film with a Cave 12½-in. $f/9$ reflector. Note the delicate shading. In the next fine photo by Dennis Milon, we see the shadow of Jupiter's satellite Ganymede in transit, just skimming the edge of an unusually darkened equatorial band. One second at $f/170$ on Royal Pan with an Astrola 8-in. reflector. Normal development in DK60a. In the last photo we see a fading out of the South Temperate Belt just above the Red Spot, a striking feature of the 1964 opposition. The dark border of the Red Spot and the central Red Spot Hollow could be seen on the original. Dennis Milon and Dewey Burkes worked together on this photo with the latter's 16-in. homebuilt Newtonian. A half second exposure on 35mm Tri-X, developed in DK60a for four minutes. The occultation of Saturn by the Moon (Fig. 10.9) was taken with an 8-in. Astrola Reflector, $f/115$ by projection, 1½ sec. on Royal Pan, developed in DK60a.

Fig. 1.7, the left mid-phases of Venus, were taken by Pfleumer with a 5½-in. $f/15$ refractor by projection to 10 times image size — 1/5 to 1/10 sec. on medium speed film. The right mid-phases were photographed by Heillegger with an 8-in. reflector and a Goodwin Barlow for an $f/38$ projection value — 1/100 sec. on Agfa Isopan Record (ASA 1,250), developed in Atomal. Extreme phases courtesy Lowell Observatory. The Mars series of Fig. 1.8 were by standard projection procedures. The two outer Jupiter photos were taken with a 12½-in. reflector, one second at

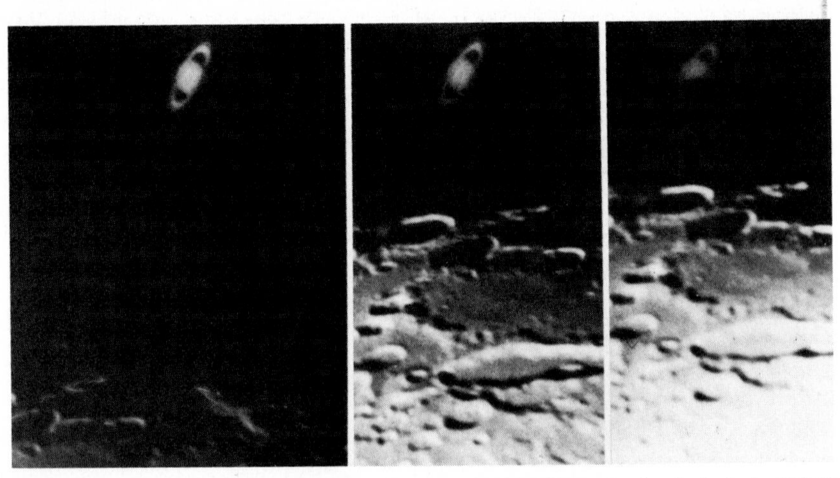

10.9 – Three phases in the occultation of Saturn by the Moon. In the last photo Saturn is partially hidden by the Moon's unilluminated invisible edge. By D. Milon with an 8-in. reflector.

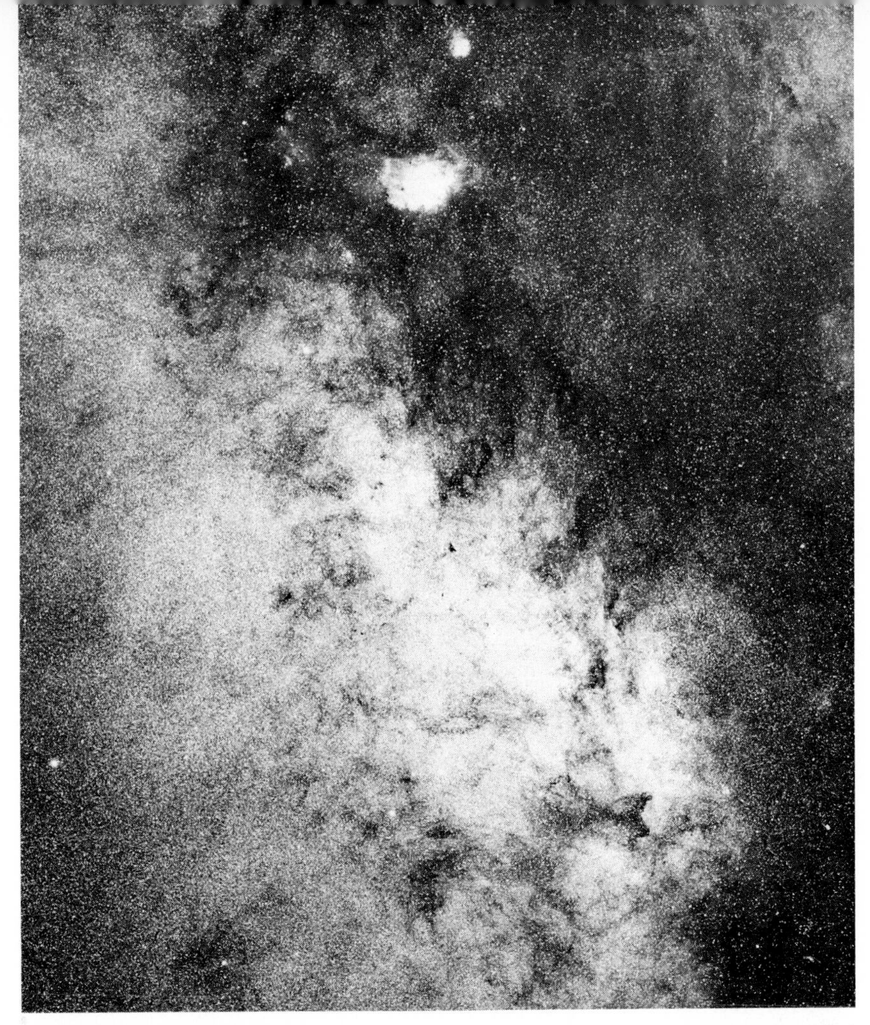

10.10 – Great masses of stars form glowing white clouds in the densely populated Sagittarius region of the Milky Way. The Lagoon nebula with the smaller Trifid above appear near the top of this fine wide-angle photo by Alan McClure. Taken with a 20-in.-focus Tessar-type lens.

f/90 on Adox KB14. The central photo, with Satellite II and its shadow, were taken with a 15½-in. Cassegrainian and projection.

In *star photography* the actual physical aperture of your telescope's lens or mirror and not its *f* value is the primary controlling factor in photographing individual stars, since stars are true point sources of light. A longer focal length will yield wider star separation on the film, but it takes larger apertures to capture fainter stars; and reciprocity law failure as described, or lack of patience, limits exposure times with smaller lenses. Figure 10.10 is a fine Milky Way Sagittarius star cloud photo taken with a 20-in.-focus Goto (Tessar type) lens at *f*/6.3, 25 minute exposure on 103aE plate through red 23A filter. Fig. 1.12, double star photos, were taken at focus of a 6-in. *f*/8 reflector, ½ and 1 second respectively on ASA 125 color film. The Milky Way was photographed with a 7-in. *f*/2.5 Aero Ektar at *f*/4, 45 minutes on Eastman 103aC emulsion. Fig. 1.13, the Pleiades was taken with 6-in. *aperture f*/3 Schmidt camera, 45 minutes on Super XX; Double Cluster, 24-in.-focus *f*/6 lens, 1½ hrs. on Tri-X through a K-3 yellow filter; the open Cluster in Cassiopea, 20-in. *aperture f*/6 Newtonian at focus, 2 hours on Super Fulgur emulsion; M-13 Hercules 12½-in. *aperture* Cave *f*/7 at focus, 30 minutes on Tri-X roll.

In *nebula photography* and other extended area work, it's the speed or *f* value that counts, much as we are familiar with in conventional photography. Most astrophotos in this area are taken at the prime focus of the lens or mirror, since we usually need all the light we can get. The famed North American nebula forms a good example since its outline can barely be envisioned with dark-adapted eyes on the darkest of nights through a maximum light-gathering binocular, as a 7 x 50. In Fig. 10.11 it has been sharply outlined by a 5-in. aperture *f*/2.0 Schmidt camera, 30 minutes on Eastman 103aE emulsion. The outstanding photo of the great Andromeda Galaxy M-31, spread across pages 138 and 139 to do it full justice, was actually taken a section at a time on three separate negatives and then combined as a photomontage and rephotographed. Each exposure was 2½ hours on Eastman 103aO plate at the prime focus of a 12½-in. *f*/8 reflector. I've seen splendid photos of Andromeda by two other outstanding amateur astrophotographers, Alan McClure and Evered Kreimer, that closely match but do not surpass this excellent photo by Dr. Custer. In Fig. 10.12 are five fine nebula and galaxy photos all at focus of a 12½-in. aperture *f*/7 Cave reflector, using Tri-X roll film. No data on exposure for the triple galaxies in Leo but for the Horsehead — 30 minutes, M-33 Spiral — 1 hour, M-27 Dumbbell — 40 minutes, and M-8 Lagoon — 40 minutes. All were developed in D-19.

Fig. 1.14, the Orion nebula, was taken with a 12½-in. *f*/7 reflector, 30 minutes on Super XX roll. An out-of-focus positive transparency was used as a dodging mask to compensate central overexposure. The ring nebula photo was taken at focus of a 12½-in. *f*/8, 2 hours on Eastman III-O plate. It was enlarged to 70X on the original print from a 1/32 inch image! In Fig. 1.15, Andromeda was photographed with a 7-in. aperture *f*/7 Fecker triplet lens, 45 minutes on a IIaO plate.

In photographing of *celestial phenomena* conditions can be extreme, and equipment must be varied accordingly. Aurora in color call for the very fastest lenses and emulsions you can commandeer. Meteor photographers prefer to use lenses at an aperture of about *f*/4, to obtain definition over a wide field and a reasonable exposure period without excess sky fog.

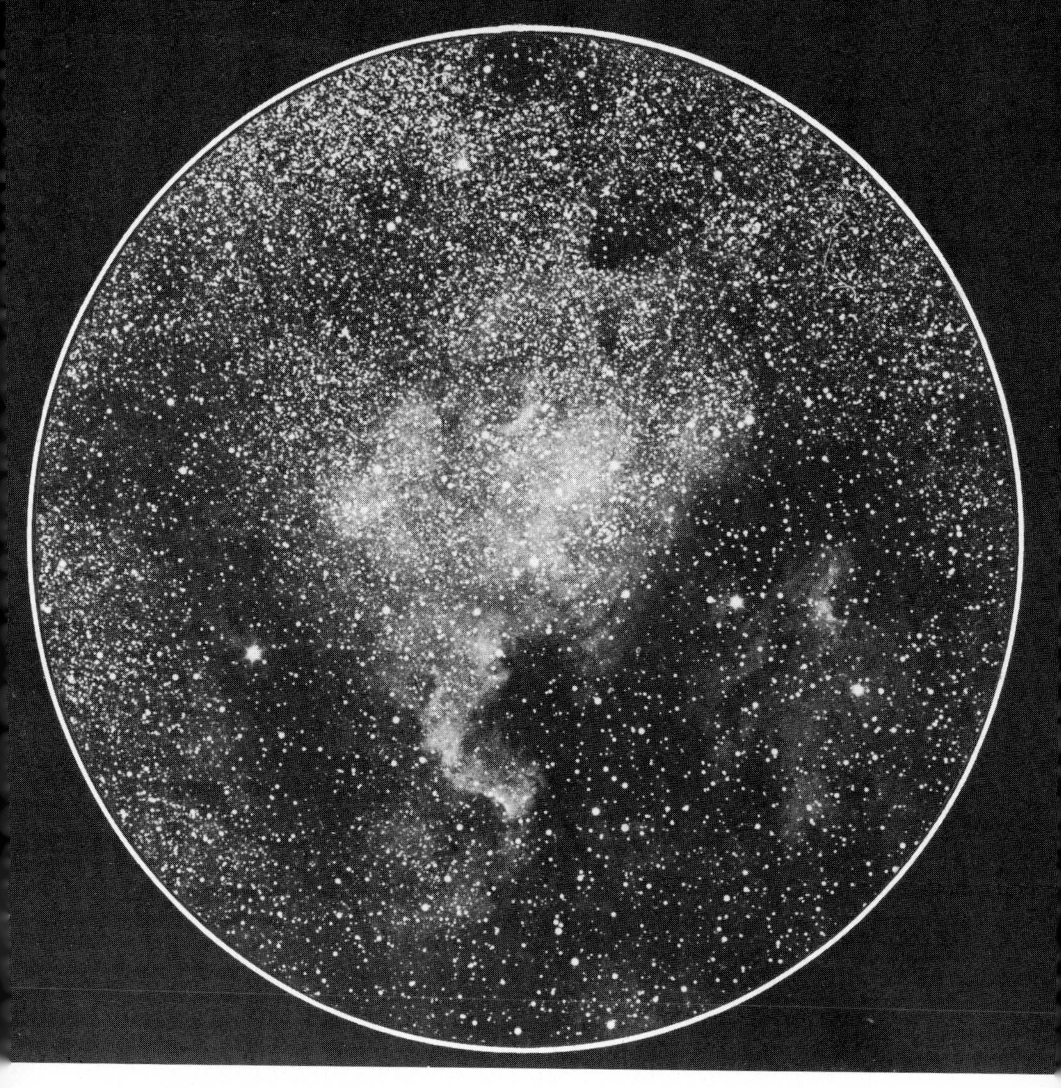

10.11 – The dim North American nebula high in Cygnus presents its true continent form in this 30 minute exposure with a 10-in. focus f/2.0 Schmidt camera lens. By the author.

Comets call for photography at the prime focus of most telescopes to utilize their full speed. Ultra-fast camera lenses should be stopped down to about $f/4$ unless the comet is too faint for this. Lunar halos require "guestimating," and for solar halos one can use the light meter on the sky away from the sun, then use this reading for maximum exposure and take several more each at a stop less. Fig. 1.16, the Perseid meteor, was caught with a 6-in. Metrogon at $f/8$ at the end of a one-hour exposure on a 103aE 8 x 10 plate; Comet Arend Roland, with a 12½-in. focus Aero-Xenar lens at $f/4.0$, 10 minutes on 103aE plate; Comet Ikeya with a 7-in.-aperture 49-in.-focus $f/7$ Fecker lens, 35 minutes on 103aO plate.

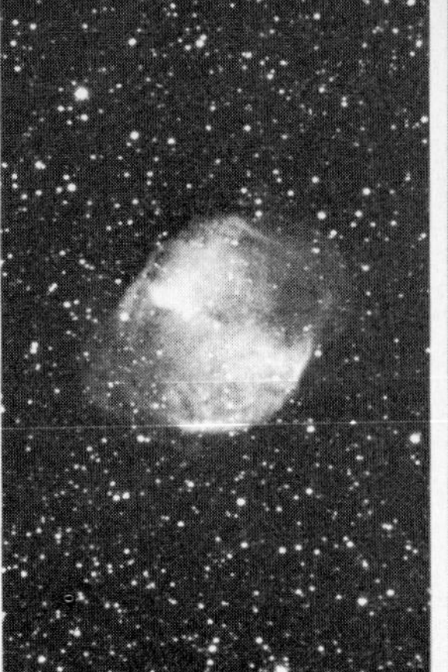

10.12 – An unusually fine photo series of some fascinating nebular wonders of the night skies; a relatively near spiral armed galaxy, M33; the famous Dumbbell nebula of Vulpecula; dark patches mottle the bright Lagoon nebula in Sagittarius. All photos by Evered Kreimer with a 12½-in. f/7 reflector.

The current edition of *Outer Space Photography* supplies information on "cold photography," and its outstanding ability to increase a film's effective speed (especially Tri-X High Speed Ektachrome), and notes on equipment required. See *Sky and Telescope* Aug. 1975 for simple construction of equipment.

Current Manufacturers Every Amateur Should Consider

This revision required that a new approach be taken to provide information on suppliers. I'm still providing names of a limited number of companies "tried and proven" over the years, parts suppliers, and some new ingenious suppliers, with the hope they will continue to survive. Listed below, alphabetically by *key* part of name, are the firms (many new entries and addresses).

AMERICAN SCIENCE CENTER (Chicago, Ill.)

AMPHOTO (Garden City, N.Y.)—publisher of my books and fine related books, specialize in photography.

ASTRONOMY CHARTED (Worcester, Mass.)—astronomical slide sets.

AUDY, G. E. (Wilmington, Del.)—excellent mirror mountings.

BURLEIGH BROOKS OPTICS INC. (Hackensack, N.J.)—cameras of all kinds, new and used, fine long-focus lenses.

BUSHNELL, D. P. & CO., INC. (Pasadena, Calif.)—a division of Bausch & Lomb Co., among the finest suppliers of binoculars and spotting scopes, also camera adapters, lenses, rifle scopes, etc.

BYERS CO., E. (Barstow, Calif.)—excellent drives and mounts.

CAVE OPTICAL CO. (Long Beach, Calif.)—large, dependable, long-established telescope makers; fine reflectors of almost any size desired; small refractors.

CELESTRON PACIFIC (Gardena, Calif.)—well-established maker of excellent Schmidt-Cassegrainian scopes, transportables 5-, 8-, and 14-in. apertures.

C. & H. SALES CO. (Pasadena, Calif.)—usable surplus items.

CLAUSING, D. L. (Skokie, Ill.)—fine Beral mirror coatings.

COULTER OPTICAL CO. (see *Sky and Telescope*)—relatively new source of *all* optics for astronomy.

CRITERION MFG. CO. (Hartford, Conn.)—established firm making fine 4- to 12-in. reflectors and an 8-in. Schmidt-Cassegrainian portable; their 6-in. reflector may be a best buy, although all are excellent.

EASTMAN KODAK CO. (Rochester, N.Y.)—best source of films and gelatin filters.

EDMUND SCIENTIFIC CO. (Barrington, N.J.)—an amateur's storehouse of optics, telescope accessories and economical scopes.

ESSENTIAL OPTICS (see *Sky and Telescope*)—8- to 18-in. reflectors plus finders on firm simple mounts at low cost.

E. & W. OPTICAL (Minneapolis, Minn.)—best reasonable source of 1/20 wave quartz or pyrex diagonals, also aluminum coatings and over-coatings.

HERBACH & RADEMAN (Philadelphia, Pa.)—excellent source of surplus motor drives and a multitude of useful items.

ILFORD (a Ciba-Geigy Co.) (Paramus, N.J.)—for films and filters.

JAEGERS, A. (Lynbrook, N.Y.)—best economical source of large (6-in. f/8-10 or 15) to small achromats plus a multitude of do-it-yourself scope parts.

LUFT, H. A. (Oakland Gardens, N.Y.)—for star atlases and books on astronomy such as *Norton's Atlas; A.T.M.* books I, II and III; etc.

MEADE INSTRUMENTS (see *Sky and Telescope*)—a relatively new company; eyepiece holders, superb guide scopes, camera adapters and do-it-yourself items.

NIKON, INC. (Ehrenreich Photo Opticals Industries, Garden City, N.Y.)—probably most versatile cameras and lenses available.

OLDEN CAMERA (New York, N.Y.)—at the crossroads of the camera business, new and used equipment.

OPTICA b/c (Oakland, Calif.)—quality supplies specifically for amateur astrophotographers and observers; an all-purpose company for kits, publications, unlimited accessories; complete scope line; four specialized catalogs.

PACIFIC INSTRUMENTS (Van Nuys, Calif.)—rings, drives, circles and mountings.

PANCRO MIRRORS INC. (Glendale, Calif.)—a highly reputable firm for optical coatings of all kinds.

PARKS, W. R. (Torrance, Calif.)—unsurpassed fiberglass telescope tubes (in colors) and mirror cells.

QUESTAR (New Hope, Pa.)—world famous 3½-in. Schmidt-Cassegrainian portable with a case size of a microscope it is a traveler's dream.

QUICK-SET INC. (Skokie, Ill.)—fine small tripods, best buys for three sizes of heavy-geared tripods for large equipment.

ROSENBERG, BEN (New York, N.Y.)—a "curio" shop for rare quality used items.

STAR-LINER CO. (Tucson, Ariz.)—reputable manufacturer of reflector telescopes, a line of economical, quality instruments.

TELESCOPTICS (Los Angeles, Calif.)—fine reflector eyepiece focusing mounts.

TUTHILL, R. W. (see *Sky and Telescope*)—a very ingenious new company; example: closed tube 4-in. Richfield reflector, perfect design, most reasonably priced; more innovations coming.

UNITRON SCIENTIFIC, INC. (Newton Highlands, Mass.)—recognized supplier of fine altazimuth, equatorial, and photo refractors; also accessories, especially refractor eyepiece mounts.

UNIVERSITY OPTICS (Ann Arbor, Mich.)—long established in economical telescopes, kits, finders and scope accessories for "do-it-yourself" builders.

VERNONSCOPE (Candor, N.Y.)—in my opinion, manufactures the world's best parfocal, wide-angle eyepieces (Brandon & Questar-Brandon); also the famous Dakin 2.4X Barlow.

Equipment and Supplies

Use the list below to find the names of various manufacturers or suppliers of the particular products you need.

ATLASES (star)—*Sky and Telescope,* Edmund, Luft, Optica b/c.

BARLOW LENSES—Vernonscope (best), Edmund, Jaegers, Telescoptics, Optica b/c.
BOOKS (astro)—*Sky and Telescope*, Luft, Amphoto, Edmund, Optica b/c, Telescoptics.
CAMERA BRACKETS (to adapt)—Optica b/c (among the best), Edmund, Criterion, Celestron, Bushnell.
CAMERAS (astro)—Criterion, Unitron, Optica b/c, Edmund, Bushnell, Celestron.
CAMERAS (with interchangeable viewfinders)—Nikon, Olympus OM-1 (compact and light weight), Canon, Miranda, Pentax (see camera journals).
CELLS (mirror)—Audy, Meade, Cave, Optica b/c.
CHARTS (rotating, etc.)—*Sky and Telescope*, Edmund, Optica b/c.
CIRCLES (setting)—Cave, Meade, Edmund, Optica b/c.
COATINGS (lens)—(aluminum and non-reflective)—Pancro (excellent), Clausing, E. & W. Optical, Miller, Woods.
CONTROLS (photo)—Edmund (low cost); good ones advertised in *Sky and Telescope*.
DIAGONALS—E. & W. Optical (best), Optica b/c, Jaegers, Unitron.
EYEPIECE HOLDERS—Telescoptics and Optica b/c among best.
EYEPIECES—Vernonscope (Brandon best), Edmund (low-cost), Jaegers, Telescoptics, Optica b/c, University, Meade.
FILM AND PLATES (astro)—Eastman Kodak, Ilford, Agfa, for small amounts; Optica b/c and offers in *Sky and Telescope*.
FILTERS—Eastman Kodak (gelatin), Nikon, Tiffon, Vivitar, Vernonscope (eyepieces).
FINDERS AND GUIDE SCOPES—Meade, Optica b/c, University Optics, most manufacturers.
LENSES (camera)—Nikon (excellent), Vivitar (very good), Canon (excellent), Leica (Leitz) (superb), Spiratone (some good buys when economy demands).
LENSES (large and small, mirrors, prisms, etc.)—Edmund, Jaegers (for large objectives), E. & W. Optical, Bausch & Lomb, C. & H. Sales.
LENSES (used-photo)—Olden, Rosenberg, Burleigh Brooks, C. & H. Sales, advertisements in *Popular Photography* and *Modern Photography*.
MAPS (star charts, etc.)—*Norton's Star Atlas* (a must), *Sky and Telescope*, Edmund, Optica b/c.
MIRRORS (cells)—Audy (best), Edmund, Cave, Meade and scope manufacturers.
MIRROR KITS—Telescoptics, Optica b/c, Precision, E. & W. Optical, Edmund, Criterion.
MIRRORS (reflectors, scopes)—Cave, Criterion, E. & W. Optical, Optica b/c, Pacific, Edmund, Coulter.
MOTORS (for driving)—Edmund (economical), E. Byers (excellent), Optica b/c, Pacific, Herbach & Rademan (surplus).
MOUNTINGS (equatorial)—Pacific, Cave, Star-Liner, Edmund, Unitron, Criterion.
OBJECTIVES (refractor)—Jaegers (the best buy for 6-in. or less, f/5 to f/15), Edmund, Cave, Rosenberg (used).
OPTICAL ELEMENTS (all kinds)—Edmund, Jaegers, Bausch & Lomb, E. & W. Optical, C. & H. Sales, Rosenberg (used).
OPTICAL GLASS—Bausch & Lomb, *Sky and Telescope*; Corning and G. E. for quartz, Cerevit, etc., new low exposure blanks.
SPIDERS—Audy, Telescoptics, Optica b/c, Edmund.
TELESCOPES (economical)—Edmund, Criterion, Jaegers (parts), Bushnell, Unitron, Sears, Montgomery Ward.
TELESCOPES (reflectors)—Cave, Essential Optics, Criterion, Star-Liner, Tuthill.
TELESCOPES (refractors)—Jaegers (best buys for lenses and parts), Cave, Edmund, Bushnell, Unitron, Criterion, Rosenberg (used).
TELESCOPES (special optics)—Questar, Celestron, Criterion (essentially Schmidt-Cassegrainian compact designs).

TELESCOPE SUPPLIES (miscellaneous)—best possible sources are advertisements in *Sky and Telescope,* Astronomy and all other journals available.

TRIPODS—Quick-Set, Edmund and war surplus outlets (see photo dealers).

TUBES—Parks and Meade for fiberglass tubes, Jaegers, Optica b/c and Edmund for aluminum.

USED INSTRUMENTS—best sources are "Sky Gazer's Exchange" in *Sky and Telescope* and "Astro-Mart" in *Astronomy,* or advertise with Rosenberg.

Literature and Books for Photographers and Observers

Every observer or builder of telescopes should subscribe to one or more of the following journals: *Sky and Telescope,* Bay State Rd., Cambridge, Mass. 02138, (caters to professionals, yet is loyal to amateurs—rich in equipment advertising); *Astronomy,* 757 N. Broadway, Suite 204, Milwaukee, Wisc. 53202, (relatively new, beautiful color outer space photos and superb artists' color renditions, latest professional information and fine practical articles for the amateur); *Modern Astronomy*, 18 Fairhaven Dr., Buffalo, N.Y. 14225, (a quarterly worth reviewing); *Astrograph,* Box 2283, Arlington, Va. 22202, (fine small bimonthly journal). Do not overlook smaller or local journals such as the *Reflector, Griffith Observer,* and so forth, which have sprung up by demand.

Regarding cameras and regular lenses see my other books and the magazines *Popular Photography* and *Modern Photography.* Always remember to check your library for the address of any publisher or journal.

Listed below are some useful books—current (C), earlier but good (E) and classic or old (O). Look for classics in libraries, used book stores or advertise for them. Please remember that astronomy and scope building are old arts and many earlier books are superb, informative and often even better than current books available.

(C) NORTON'S STAR ATLAS by Norton and Inglis (Sky Pub. Corp.)—an absolute must for every photographer and observer, 16th edition just out, from *Sky and Telescope,* Luft or Edmund.

(C) THE AMATEUR ASTRONOMER'S HANDBOOK—1974 ed. by J. Muirden (T.Y. Crowell Co.)—This British author's fine book covers a little of everything; a useful reference book.

(C) ALL ABOUT TELESCOPES by Sam Brown (Edmund Scientific Co.)—for beginners, available in paperback from Edmund.

(E) ASTRONOMICAL PHOTOGRAPHY AT THE TELESCOPE by T. Rackham (Macmillan)—for 6-in. telescopes and larger, not recent but good, from *Sky and Telescope* and Edmund.

(E) SKYSHOOTING by Mayall and Mayall (Ronald Press)—excellent for beginners.

(E) AMATEUR TELESCOPE MAKING ed. by A.G. Ingalls (Scientific American, Inc.)—Books I, II, III, though approximately $10 ea., are a gold mine despite little or no indexing; beginners buy I from Luft, Edmund, *Sky and Telescope.*

(E) HOW TO MAKE A TELESCOPE by Texerau (Interscience Pub.)—outstanding for basic principles, involved yet readable, from Edmund or *Sky and Telescope.*

(E) MAKING YOUR OWN TELESCOPE by Thompson (Sky Pub. Corp.)—is the old "stand by" for detailed instructions, from *Sky and Telescope.*

(O) CELESTIAL OBJECTS FOR COMMON TELESCOPES (2 Vol.) by Webb (paperback now available through Dover Publishing Co.)—original volumes now collector's items—available from Edmund.

(O) THE TELESCOPE by L. Bell (McGraw-Hill)—very rare but filled with practical information and history, should be reprinted.

(O) THE HISTORY OF THE TELESCOPE by H.C. King (Chas. Griffin & Co. Ltd.) —The most complete history of telescopes every written—rare.

REGARDING JOURNALS—obtain those suggested or visit your local library and review carefully—especially available back issues. My older third edition (1967) of OUTER SPACE PHOTOGRAPHY has a fine review of earlier articles.

INDEX